Springer Series in
MATERIALS SCIENCE 141

Springer Series in
MATERIALS SCIENCE

Editors: R. Hull   C. Jagadish   R.M. Osgood, Jr.   J. Parisi   Z. Wang   H. Warlimont

The Springer Series in Materials Science covers the complete spectrum of materials physics, including fundamental principles, physical properties, materials theory and design. Recognizing the increasing importance of materials science in future device technologies, the book titles in this series reflect the state-of-the-art in understanding and controlling the structure and properties of all important classes of materials.

Please view available titles in *Springer Series in Materials Science*
on series homepage http://www.springer.com/series/856

Miki Niwa
Naonobu Katada
Kazu Okumura

Characterization and Design of Zeolite Catalysts

Solid Acidity, Shape Selectivity and Loading Properties

With 135 Figures

Prof. Miki Niwa
Tottori University
Graduate School of Engineering
Dept. Chemistry & Biotechnology
Koyama-Minami 4-101
680-8552 Tottori, Japan
mikiniwa@chem.tottori-u.ac.jp

Prof. Naonobu Katada
Tottori University
Graduate School of Engineering
Dept. Chemistry & Biotechnology
Koyama-Minami 4-101
680-8552 Tottori, Japan
katada@chem.tottori-u.ac.jp

Prof. Kazu Okumura
Tottori University
Graduate School of Engineering
Dept. Chemistry & Biotechnology
Koyama-Minami 4-101
680-8552 Tottori, Japan
okmr@chem.tottori-u.ac.jp

Series Editors:

Professor Robert Hull
University of Virginia
Dept. of Materials Science and Engineering
Thornton Hall
Charlottesville, VA 22903-2442, USA

Professor Jürgen Parisi
Universität Oldenburg, Fachbereich Physik
Abt. Energie- und Halbleiterforschung
Carl-von-Ossietzky-Straße 9–11
26129 Oldenburg, Germany

Professor Chennupati Jagadish
Australian National University
Research School of Physics and Engineering
J4-22, Carver Building
Canberra ACT 0200, Australia

Dr. Zhiming Wang
University of Arkansas
Department of Physics
835 W. Dicknson St.
Fayetteville, AR 72701, USA

Professor R.M. Osgood, Jr.
Microelectronics Science Laboratory
Department of Electrical Engineering
Columbia University
Seeley W. Mudd Building
New York, NY 10027, USA

Professor Hans Warlimont
DSL Dresden Material-Innovation GmbH
Pirnaer Landstr. 176
01257 Dresden, Germany

Springer Series in Materials Science ISSN 0933-033X
ISBN 978-3-642-12619-2   e-ISBN 978-3-642-12620-8
DOI 10.1007/978-3-642-12620-8
Springer Heidelberg Dordrecht London New York

Library of Congress Control Number: 2010933957

© Springer-Verlag Berlin Heidelberg 2010
This work is subject to copyright. All rights are reserved, whether the whole or part of the material is concerned, specifically the rights of translation, reprinting, reuse of illustrations, recitation, broadcasting, reproduction on microfilm or in any other way, and storage in data banks. Duplication of this publication or parts thereof is permitted only under the provisions of the German Copyright Law of September 9, 1965, in its current version, and permission for use must always be obtained from Springer. Violations are liable to prosecution under the German Copyright Law.

The use of general descriptive names, registered names, trademarks, etc. in this publication does not imply, even in the absence of a specific statement, that such names are exempt from the relevant protective laws and regulations and therefore free for general use.

Cover design: SPi Publisher Services

Printed on acid-free paper

Springer is part of Springer Science+Business Media (www.springer.com)

Preface

Catalysis and catalyst is a key technology to solve the problems in energy and environment issues to sustain our human society. We believe that comprehensive understanding of the catalysis and catalyst provides us a chance to develop a new catalyst and contributes greatly to our society. However, the field of heterogeneous catalyst is difficult to study and still stays behind more developed fields of chemistry such as organic and physical chemistries. This is a dilemma to the chemists who study the catalysis and catalyst. While we can accomplish the progress in the industrial application, the scientific understanding is not complete yet. A gap between the useful application and incomplete scientific understanding, however, becomes smaller and smaller in recent years.

Because zeolites are fine crystals, and the structure is clearly known, the study on the catalysis using the zeolites is easier than those encountered in other catalysts such as metals and metal oxides. Very fortunately, zeolites provide us the strong acidity with the fine distribution which enables various useful catalytic reactions. When some metals and cations are loaded in close to the acid sites, these loaded elements show extraordinary characters, and many catalytic reactions proceed thereon. Microporosity provides us a unique property of the shape selectivity, and the excellent selectivity is observed. These various beneficial properties of zeolites lead us to study the fundamental chemistry which is related to the catalytic properties. In this book, we want to summarize three important properties of zeolites: solid acidity, shape selectivity, and loading property. To accomplish the study on them, three kinds of important techniques are explained in detail, i.e., temperature-programmed desorption of ammonia, chemical vapor deposition of silica, and EXAFS measurement. We will focus on our interests on these characterization techniques and introduce the reader of this book to our characterization school for zeolite catalysts.

Three of us are staffs in Tottori University, Japan, and education and research are our duty. It is our pleasure that we study and learn the chemistry of catalysis by collaborating with many students in this laboratory. Our study depends greatly on the collaboration with them.

We first proposed the publication of this book to Mr. Ippei Ohta belonging to Tokyo Institute of Technology Press, but unfortunately this publication was not related to his recent job. However, he introduced to us Dr. Claus E. Ascheron,

Springer Science and Business Media, Heidelberg, Germany; and he accepted our proposal. Without his strong support to our activity, this book would never be published. We want to express our sincere thanks to Mr. Ohta and Dr. Ascheron.

Tottori, Japan *Miki Niwa*
May 2010 *Naonobu Katada*
Kazu Okumura

Contents

Preface v

1 Introduction to Zeolite Science and Catalysis 1
   1.1 General Trend of the Zeolite Acid Catalyst: Species,
        Structure, and Industrial Application 1
   1.2 Property of Zeolite Catalyst: Acidity, Shape Selectivity,
        and Loading Property 5
   1.3 Intention of the Publication with the Background
        of the Zeolite Chemistry 6
   References 8

2 Solid Acidity of Zeolites 9
   2.1 Multiple Characterization Techniques 9
        2.1.1 Physical Properties 9
        2.1.2 Chemical Properties 10
   2.2 Fundamentals for the TPD of Ammonia 11
        2.2.1 Experimental Apparatus of the TPD Method 11
        2.2.2 Identification of the Desorption, l- and h-Peaks 13
        2.2.3 Identification of Ammonia Desorbed from Y Zeolite 14
   2.3 Theory for the TPD of Ammonia 15
        2.3.1 Conditions of the Equilibrium 15
        2.3.2 Derivation of a Fundamental Equation 17
        2.3.3 Determination of ΔH and Constancy of ΔS (Desorption) 19
   2.4 Practical Measurements of Ammonia TPD 20
        2.4.1 Curve Fitting Method to Measure the Strength of Acid Site ... 20
        2.4.2 ΔH on Various Zeolites with Different
                Concentrations of Acid Site 21
   2.5 Number of Acid Sites on Various Zeolites 24
        2.5.1 Number of Acid Sites Correlated with Al Concentration 24
        2.5.2 In Situ and Ex Situ Prepared H-type Zeolites 26
   References 27

3 IRMS-TPD Measurements of Acid Sites ... 29
3.1 Measurement Method ... 29
3.1.1 What Is Obtained from the TPD Measurement ... 29
3.1.2 Experimental Methods ... 29
3.1.3 Required Corrections, IR Band Position and ΔS ... 31
3.2 Proton Form Zeolite ... 32
3.2.1 H-Mordenite ... 32
3.2.2 H-Y and H-Chabazite ... 34
3.3 Modified Zeolites ... 39
3.3.1 Multivalent Cation-Modified Zeolite ... 39
3.3.2 Ultrastable Y (USY) Zeolite ... 41
3.4 Distribution of Brønsted Acid Sites Dependent on the Concentration ... 43
3.5 Relationship Between Stretching Frequency and Ammonia Desorption Heat of OH Group ... 45
3.6 Distorted Structure of Zeolite and Related Material with Lewis Acidity and Broad Distribution of Acid Strength ... 47
3.7 Measurements of Metal Oxide Overlayer ... 52
3.8 Extinction Coefficients of NH_4^+ and NH_3 Adsorbed on Brønsted and Lewis Acid Sites, Respectively ... 55
References ... 59

4 DFT Calculation of the Solid Acidity ... 61
4.1 DFT Calculation ... 61
4.1.1 DFT Calculation Applied to the Study on Brønsted Acidity ... 61
4.1.2 Embedded Cluster and Periodic Boundary Conditions ... 62
4.2 Application to Chabazite, a Simple Zeolite ... 63
4.2.1 Brønsted Acid Sites in Chabazite Based on the Models Within the Periodic Boundary Conditions ... 63
4.2.2 Brønsted Acid Site in an Embedded Cluster Model ... 65
4.3 Application to Other Zeolites ... 65
4.3.1 FAU, MOR, and BEA Calculated Under the Conditions of the Embedded Cluster and the Periodic Boundary ... 65
4.3.2 MFI, FER, and MWW Calculated Under the Embedded Cluster Model ... 67
4.4 Modified Zeolites ... 69
4.4.1 Divalent Cation-Exchanged Y Zeolites Based on the Embedded Cluster Model ... 69
4.4.2 Modified Brønsted OH in Y Zeolite Based on the Periodic Boundary Conditions ... 71
4.5 Dependence of Brønsted Acid Strength on Local Geometry ... 72
References ... 78

5	**Catalytic Activity and Adsorption Property**		79
	5.1 Paraffin Cracking		79
		5.1.1 Evaluation of Intrinsic Activity of Acid Site	79
		5.1.2 Dependence of Activity on Acid Strength	81
		5.1.3 Thermodynamic Description on Correlation Between Activation Energy and Ammonia Desorption Heat	86
		5.1.4 Behavior of Acid Sites in 8- and 12-Rings of Mordenite	88
	5.2 Adsorption of Aromatic Hydrocarbons		89
	5.3 Friedel–Crafts Alkylation on Ga-MCM-41		93
	5.4 Amination of Phenol into Aniline on Ga/ZSM-5		97
	References		100
6	**CVD of Silica for the Shape Selective Reaction**		103
	6.1 Reactants and Products Shape Selectivity, Concept and Definition		103
	6.2 Chemical Vapor Deposition of Silica and the Procedure		104
	6.3 Formation of Silica Overlayer on the External Surface		108
		6.3.1 Method of Benzene-Filled Pore for the Measurement of External Surface Area	108
		6.3.2 Mechanism of CVD to form the Silica Overlayer	109
		6.3.3 Formation of Silica Overlayer on Zeolite and Metal Oxide, and Its Function	110
	6.4 Fine Control of Pore-Opening Size		111
		6.4.1 Mordenite	112
		6.4.2 MFI Zeolite	113
		6.4.3 A Zeolite	114
		6.4.4 Y Zeolite	119
	6.5 Characterization of Deposited Oxide		120
		6.5.1 XPS Measurements	120
		6.5.2 EXAFS of the Deposited Germanium	121
		6.5.3 TEM Observation	123
	6.6 External Surface Acidity: Measurements and Inactivation		124
	References		127
7	**Application of the CVD of Silica to the Shape Selective Reaction**		129
	7.1 Selective Formation of *Para*-Dialkylbenzene		129
		7.1.1 Principle of the Shape Selectivity	129
		7.1.2 CVD Zeolite to Produce the Para-Dialkylbenzene	131
		7.1.3 In Situ Production of CVD Zeolites	135
		7.1.4 HZSM-5 In Situ and Ex Situ Prepared for the CVD of Silicon Alkoxide	136
		7.1.5 CLD of Silica for the Shape Selective Adsorption	136
	7.2 Selective Cracking of Linear Alkane (Dewaxing)		139

7.3 Various Applications ... 141
    7.3.1 Preferential Production of Dimethylamine from Methanol and Ammonia ... 141
    7.3.2 Improvement of the Life and the Activity of Catalysts ... 142
    7.3.3 Selective Removal of Undesired Products ... 144
    7.3.4 Applications to Zeolites from Various View-Points ... 144
References ... 146

8 Zeolite Loading Property for Active Sites and XAFS Measurements ... 149
8.1 EXAFS and XANES Measurements of Loaded Metals ... 149
    8.1.1 DXAFS and QXAFS Analysis ... 149
    8.1.2 Formation of Molecular-Like PdO Through the Interaction with Acid Sites of Zeolites ... 150
    8.1.3 Reversible Cluster Formation Through the Interaction with Acid Sites of Zeolites ... 152
8.2 In Situ QXAFS Studies on the Dynamic Coalescence and Dispersion Processes of Pd in USY Zeolite ... 154
8.3 Formation of the Atomically Dispersed Pd^0 Through H_2 Bubbling in o-Xylene: XAFS Measurements of Metals in the Liquid . 157
References ... 162

9 Catalytic Reaction on the Palladium-Loaded Zeolites ... 163
9.1 Combustion of Hydrocarbons Over Pd-Supported Catalysts ... 163
    9.1.1 Toluene Combustion ... 164
    9.1.2 Methane Combustion ... 165
9.2 Selective Reduction of NO with Methane in the Presence of Oxygen . 169
    9.2.1 Improvement in the Activity Derived by the Combined Effect of Adsorbent of Aromatic Acids ... 170
9.3 Cross-Coupling Reactions Over Pd Loaded on FAU-Type Zeolites ... 172
    9.3.1 Heck Coupling Reactions Over Pd Loaded on H-Y Zeolites ... 172
    9.3.2 Remarkable Enhancement of Catalytic Activity Induced by the H_2 Bubbling in Suzuki–Miyaura Coupling Reactions ... 173
    9.3.3 A Possible Mechanism for the Formation of Active Pd Species in o-Xylene ... 176
References ... 178

Index ... 181

Chapter 1
Introduction to Zeolite Science and Catalysis

Abstract General trend of the zeolite acid catalyst is explained as an introduction. Species, structure, and industrial application of the zeolites are explained. Function of zeolites as a catalyst is based on three important properties: solid acidity, shape selectivity, and loading property. The purpose of the chapter is, therefore, not only to review the important functions of zeolite catalyst, but also to explain the application of these properties to the developed zeolite catalysts.

1.1 General Trend of the Zeolite Acid Catalyst: Species, Structure, and Industrial Application

Zeolite is a porous crystal typically consisting of Si, Al, and O atoms, and a catalytic material with vast industrial applications. From the early application to petroleum refinery into the recent utilization in the green sustainable chemical processes, many kinds of zeolites are developed and applied for the industrial applications.

About 190 kinds of framework type codes have been given to only known structures of zeolites, zeolitic silicates and phosphates to date (2010), and every year the number of zeolite species is increasing, according to the International Zeolite Association Web site [1]. Since the remarkable developments of the mesoporous materials such as MCM-41 and FSM-16, many kinds of mesoporous materials also are synthesized and studied. Therefore, many opportunities to develop a new catalyst based on these porous materials are given.

However, actually, there are only about ten kinds of zeolites which have been applied to the industrial processes, as shown in Table 1.1. Because the thermal and mechanical stabilities are not enough to use the industrial process, and the zeolite synthesis requires too much cost and time, so many kinds of zeolite species are not available industrially. Therefore, Y-zeolite, ZSM-5, mordenite, MCM-22, and β zeolites are most typical zeolite catalysts. These are modified adequately, for example, by steaming, HCl dealumination, cation-exchange, and metal loading to be utilized as industrial catalysts. Almost all the processes listed in Table 1.1 are acid-catalyzed reactions only except for H_2O_2 oxidation on titanosilicate. In other words, the catalytic processes utilized for the zeolite species are associated with catalysis

Table 1.1 Zeolite catalysts applied for the industrial processes and the environmental protection

Zeolite species		Process	Company
Y (FAU)	USY, CaHY, LaHY	Catalytic cracking	
ZSM-5 (MFI)		Isomerization	
		Alkylation	
		Disproportionation	
	Silicalite	Beckmann rearrangement [2]	Sumitomo Chemical
	Me-MFI	Pyridine bases synthesis [3]	Koei Chemical
	(dispersed)	Hydration of cyclohexene [4]	Asahi Chemical
		Diethanolamine synthesis [5]	Nippon Shokubai
	Me(Cu [6], Fe [7], Co [8])-MFI	NO reduction	
	Mo-MFI [9]	Aromatization of methane	
Mordenite (MOR)		Trans-alkylation, Isomerization, Disproportionation	
	(modified)	Methylamine synthesis [10]	Nitto Chemical
MCM-22 (MWW)		Alkylation of benzene [11]	Mobil
β (BEA)		Alkylation of benzene	
		Acylation with acetic anhydride [12]	
L (LTL)	Pt-L	Aromatization [13, 14]	
SAPO-34		Methanol to olefin	UOP
Titanosilicate		H_2O_2 oxidation [15, 16]	Eni-Chem

on acid sites. Strong acidity with a fine distribution must be the most important property observed in the zeolite catalyst.

Structures of these typical zeolites are shown in Figs. 1.1–1.5, in which the micropore is clearly observed, and the details of the pore channel system are described in Table 1.2 [17]. Structure of zeolites is one (MOR and FER), two (MWW), or three (FAU, MFI, CHA, and LTA) dimensional. Two channels are interconnected (MFI, MOR, *BEA, and FER), or separated (MWW). In FAU, CHA, and LTA, one kind of channel forms the pore in the three-dimensional structure, and the structure of FAU is shown in Fig. 1.1. MFI has the interconnected channels parallel to [100] and [010] to form the three-dimensional structure, as shown in Figs. 1.2 and 1.3, respectively. MOR shows the outstanding one-dimensional tube of 12-ring (12 oxygen-membered ring) shown in Fig. 1.4, but another channel is interconnected to make the so-called side pocket of 8-ring shown in Fig. 1.5. Beta zeolite is shown by *BEA with the asterisk, because it is not consisted of a pure crystal phase but three polymorphs [18].

The micropore of zeolite is typically constructed from 12-, 10- and 8-rings, and sizes of the pores are roughly 0.7 nm, 0.55 nm, and 0.4 nm, respectively. The shape and size of pore, which is similar to the size of small molecules, is a fundamental reason of the shape selectivity. Zeolites are thus clearly discriminated from other kinds of catalytic materials based on the shape selectivity.

1.1 General Trend of the Zeolite Acid Catalyst: Species, Structure, and Industrial Application 3

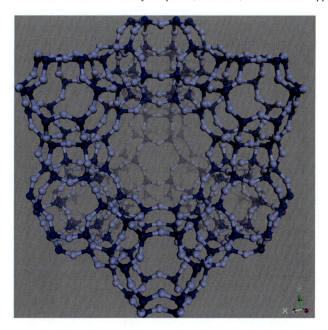

Fig. 1.1 FAU structure

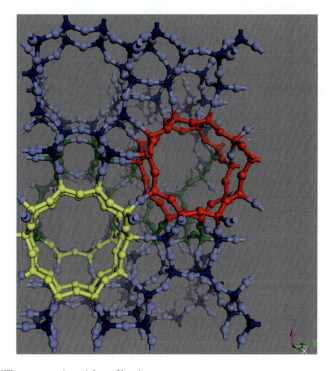

Fig. 1.2 MFI structure viewed from X-axis

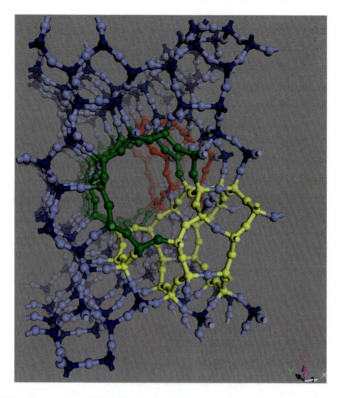

Fig. 1.3 MFI structure viewed from Y-axis

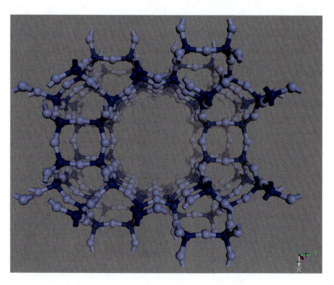

Fig. 1.4 MOR structure viewed from Z-axis

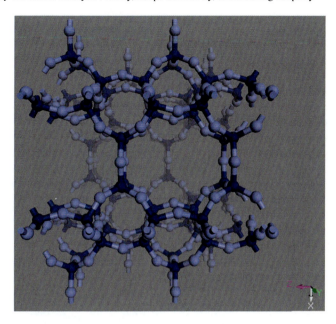

Fig. 1.5 MOR structure viewed from Y-axis

Table 1.2 Zeolite channel system

Zeolite name	Framework type code	Channels[a]	T-site[b]
Y	FAU	<111> **12** 7.4 × 7.4***	1
ZSM-5	MFI	{[100] **10** 5.1 × 5.5 ↔ [010] **10** 5.3 × 5.6}***	12
Mordenite	MOR	[001] **12** 6.5 × 7.0* ↔ [001] **8** 2.6 × 5.7***	4
MCM-22	MWW	⊥[001] **10** 4.0 × 5.5** | ⊥[001] **10** 4.1 × 5.1**	8
β	*BEA	<100> **12** 6.6 × 6.7** ↔ [001]**12** 5.6 × 5.6*	9
Chabazite	CHA	⊥[001] **8** 3.8 × 3.8***	1
Ferrierite	FER	[001] **10** 4.2 × 5.4* ↔ [010] **8** 3.5 × 4.8*	4
A	LTA	<100> **8** 4.1 × 4.1***	1

[a]Number of asterisk (*) means that the structure is one, two, or three-dimensional. Bold type character **8**, **10**, or **12** means number of oxygen to form the pore ring which has the sizes shown by $a \times b$ in Å. Two channels in zeolites MFI, MOR, MWW, *BEA, and FER are interconnected (↔) or separated (|). The channel is viewed parallel to a specific direction or not; and the former is shown by, e.g. [100], and the latter by, e.g. <100>. Rings in MWW and CHA are viewed normal (⊥) to [001]
[b]Number of non-equivalent T-site

1.2 Property of Zeolite Catalyst: Acidity, Shape Selectivity, and Loading Property

Functions of the zeolite as catalytic materials are based on three important properties, strong acidity, shape selectivity, and loading property, and new zeolite catalysts have been developed utilizing these properties effectively.

Zeolite is one of the metal oxides which have the strong solid acidity. The strong acidity in zeolites is generated by aluminum atoms incorporated into the T-site (tetrahedral site) of the framework, and the strength of the acidity depends on the microstructure of zeolite framework. On the other hand, metal oxides do not have such a fine and sharp distribution of the acid strength. In many cases, the generation of acid sites on metal oxides is not known unambiguously. Fine property of the zeolite acidity, therefore, leads us to the design of active and selective catalysts.

Among zeolite catalysts applied to the industrial processes, ultra stable Y (USY) contributes greatly to our human society, because it plays a role of the active catalyst working in the petroleum refinery process for a long time. Catalytic cracking of the petroleum makes a fundamental basis of the global economy. However, chemical processes do not require the strong acidity always, but the weak acidity may be enough. The number of acid site is controlled by decreasing the content of aluminum in the framework, for example, in order to elongate the life of catalyst. Zeolite species have the acidity, of which strength is weak to strong in the distribution sequence, Y < β < ZSM-5 ≈ MCM-22 < mordenite. A zeolite catalyst is selected from the species depending on the desired catalytic activity.

Shape selectivity is a unique property of the zeolite. Catalytic reactions are controlled by utilizing the property, and the high selectivity of reactions could be achieved. Selectivity due to the molecular sizes of reactant or product is a concept only available for the zeolite catalyst. However, as mentioned above, there are not so many kinds of zeolites available for use. Adjusting the pore size is therefore required to have the fine shape selectivity. Chemical vapor deposition of silica on the external surface of zeolites is one of the methods to effectively control the pore-opening size without affecting the internal solid acidity.

Zeolites have a high surface area, which is ca. $400 \, m^2 \, g^{-1}$. Because the composition of Al and Si atoms in the framework can be controlled, the overall property of acidity and basicity of the zeolite surface is controlled. The high surface area and the controllable acid and base property are the fundamental reasons why the zeolite is utilized as a support of metals and metal oxides. Loaded metals and cations are affected by the zeolite structure to become electron-rich or poor in nature. Such an alteration could induce the new catalytic activity. Thus, an extremely high dispersion of loaded metal or metal oxide can be achieved in the interior of zeolites.

1.3 Intention of the Publication with the Background of the Zeolite Chemistry

The purpose of the chapter is, therefore, not only to review the important functions of zeolite catalyst, but also to explain the application of these properties to the developed zeolite catalysts.

Characterizations of the zeolite acidity are extremely important, and understanding of the acidic property is necessary to develop new types of zeolite catalyst. Therefore, the investigation on the zeolite acidity is the first topic of this chapter.

1.3 Intention of the Publication with the Background of the Zeolite Chemistry

We will summarize our own investigation on the zeolite acidity by means of a method of temperature-programmed desorption (TPD) of ammonia. Fundamentals and utilization of the TPD experiment are explained in detail. Practical measurement methods to determine number and strength of the zeolite acidity are introduced, from which even the beginners can measure the solid acidity precisely. As an advanced method, infrared spectroscopy/mass spectroscopy combined method of ammonia TPD (IRMS-TPD) is summarized. By means of this method, the structure of acid site (Brønsted and Lewis acid sites) is studied, and the distribution of the Brønsted acid sites is revealed. In addition, theoretical study of density functional theory (DFT) calculation is introduced, and further combining with the ammonia TPD measurements provides us an elegant tool for the characterization of acid sites. Based on these measurements, the zeolite acidity is precisely explained in an atomic level.

Relationship between the solid acidity and catalytic activity is so important to the researchers in this field of catalysis investigation. Catalytic cracking of hydrocarbons on the Brønsted acid site and Friedel–Crafts alkylation on the Lewis acid site are explained based on the characterized zeolite acidity.

Property of the zeolite, which is discriminated from other kinds of catalysts, is the shape-selectivity. This is based on the molecular sieving property of zeolites. Utilization of the microporosity is benefit for the industrial application, because the energy utilization is optimized. We propose a method of chemical vapor deposition (CVD) of silicon alkoxide to control the pore-opening size precisely, because the obtained pore size is insufficient to realize the 100% selectivity of the catalytic reaction. Using the CVD technique, an extremely high selectivity to produce p-xylene as a result of methylation or disproportionation of toluene is achieved. Preparation of the CVD zeolite, mechanism of the deposition of silica, and selectivity of p-xylene formation are mentioned precisely. This study is applied to many other kinds of zeolite, and the separation of small molecules is made possible. Chemical liquid deposition (CLD) of silicon alkoxide is also proposed for the same purpose, and it will be compared with the CVD technique.

Metal-loaded zeolites are utilized for multiple purposes, e.g., oxidation, hydrogenation, organic reactions, etc. Activity and selectivity depend on the loaded species and their stabilized positions. Among various methods of characterization, EXAFS (extended X-ray absorption fine spectrum) measurements are recently known to be the most powerful technique. Fundamentals of EXAFS and XANES (X-ray absorption near edge spectrum) measurements of the loaded Pd are elucidated. Based on these measurements, Pd metal and PdO formed in the exterior and interior of the zeolite, respectively, are revealed. The utilization of the characterized data toward understanding the catalytic activity will be mentioned exemplified with the reactions of combustion of methane and selective reduction of NO with methane in the presence of oxygen. Pd nano cluster is stabilized in the Y zeolite cavity, and the size of the Pd cluster is adjusted by repetition of oxidation and reduction cycles. In particular, an extremely high catalytic activity of the Pd monoatomic species stabilized in the super cage of Y zeolite for the Suzuki–Miyaura reaction is described.

Heterogeneous catalysis is a key technology which enables us to realize the green sustainable chemical process. Energy saving, low cost to the environment, and utilization of a small amount of precious element are required to do the green chemistry. The main purpose of this book is, therefore, to introduce the researcher in the field of zeolite chemistry to develop the new environment benign chemical process based on the utilization of zeolite catalysts.

References

1. Ch. Baerlocher, L.B. McCusker, Database of zeolite structures. http://www.iza-structure.org/databases/
2. H. Ichihashi, M. Ishida, A. Shiga, M. Kitamura, T. Suzuki, K. Suenobu, K. Sugita, Catal. Surv. Asia **7**, 261 (2003)
3. S. Shimizu, N. Abe, A. Iguchi, H. Sato, Catal. Surv. Jpn. **2**, 71 (1998)
4. H. Ishida, Catal. Surv. Jpn. **1**, 241 (1997)
5. H. Tsuneki, M. Kirishiki, T. Oku, Bull. Chem. Soc. Jpn. **80**, 1075 (2007)
6. M. Iwamoto, H. Yahiro, K. Tanda, N. Mizuno, Y. Mine, S. Kagawa, J. Phys. Chem. **95**, 3727 (1991)
7. X.B. Feng, W.K. Hall, J. Catal. **166**, 368 (1997)
8. Y.J. Li, J.N. Armor, Appl. Catal. B Environ. **1**, L31 (1992)
9. D.J. Wang, J.H. Lunsford, M.P. Rosynek, J. Catal. **169**, 347 (1997)
10. Y. Ashina, T. Fujita, M. Fukatsu, K. Niwa, J. Yagi, in *Proceedings of the 7th International Zeolite Conference* (Kodansha/Elsevier, Tokyo/Amsterdam, 1986), p. 779
11. A. Corma, V. Martinez-Soria, E. Schnoeveld, J. Catal. **192**, 163 (2000)
12. U. Freese, F. Heinrich, F. Roessner, Catal. Today **49**, 237 (1999)
13. T.R. Hughes, W.C. Buss, P.T. Tamm, R.L. Jacobson, in *Proceedings of the 7th International Zeolite Conference* (Kodansha/Elsevier, Tokyo/Amsterdam, 1986), p. 725
14. R.E. Jentoft, M. Tsapatsis, M.E. Davis, B.C. Gates, J. Catal. **179**, 565 (1998)
15. M. Taramasso, G. Perego, B. Notari, U.S. Patent 4 410 501 (1983)
16. T. Tatsumi, M. Nakamura, S. Negishi, H. Tominaga, J. Chem. Soc. Chem. Commun. 476 (1990)
17. Ch. Baerlocher, L.B. McCusker, D.H. Olson (eds.), *Atlas of Zeolite Framework Types*, 6th Revised edn. (Elsevier, Amsterdam, 2007)
18. J.B. Higgins, R.B. LaPierre, J.L. Schlenker, A.C. Rohrman, J.D. Wood, G.T. Kerr, W.J. Rohrbaugh, Zeolites **8**, 446 (1988)

Chapter 2
Solid Acidity of Zeolites

Abstract Principle and method of the ammonia temperature-programmed desorption (TPD) are introduced to characterize the zeolite acidity. Quantitative measurements of number and strength of the acid sites on zeolites using the ammonia TPD are described. Number and strength on zeolites thus measured are well correlated with the content of Al in the framework and the structure of zeolites, respectively.

2.1 Multiple Characterization Techniques

Fundamental characterization techniques are all inevitable to study the zeolite property and catalytic activity. However, depending on the research conditions, some of them can be skipped. Because zeolite is a crystal material with a microporosity, following physical and chemical characterization techniques are usually required.

2.1.1 Physical Properties

X-ray diffraction (XRD) is measured for identification of the crystal phase. Simultaneously, the distance between lattice planes is measured using the Bragg equation. From the lattice constant, the concentration of Al in the zeolite framework is estimated [1]. The degree of crystalline formation is estimated from the diffraction intensity.

Scanning electron microscopy (SEM) is taken to see the crystal morphology, and used for a direct measurement of the crystal size. However, it is often difficult to identify whether the observed grain consists of a true crystalline. Transmission electron spectroscopy (TEM) is also utilized to show the crystal morphology, content of crystallized phase in a particle, and preferentially grown plane. TEM is, in addition, only a method that can directly visualize the microporous structure. Recently, TEM is combined with electron diffraction to determine the structure of new zeolite framework [2].

Magic angle spinning nuclear magnetic resonance (MAS NMR) is measured to distinguish between tetrahedral and octahedral configurations of ^{27}Al, and in some cases, to analyze the detailed microstructure of each species. In addition, the environmental conditions of ^{29}Si are measured mainly to estimate the framework Al concentration and the amount of SiOH [3]. The acidic OH group can be observed directly by ^{1}H-NMR, if sufficiently high resolution is obtained [4].

X-ray photoelectron spectroscopy (XPS) of Al is measured to identify the distribution of Al along the depth of zeolite crystal. When the Al concentration as measured from XPS is larger than the chemical composition, Al atom is enriched on the external surface; at the conditions, the acid sites are not uniformly distributed in the zeolite crystal to be enriched on the external surface.

Adsorption isotherm of nitrogen is measured to study the microporosity. Type I of the nitrogen isotherm is a proof of the microporosity. Deviation from the type I isotherm indicates the presence of amorphous materials mixed with the zeolite crystal. The degree of crystalline formation can be estimated from micropore volume. Mesoporosity is also known from the type IV isotherm.

A method of benzene-filled pore, namely, nitrogen adsorption after filling the micropore with benzene is used for measurements of the external surface area [5], as shown in Chap. 6.

2.1.2 Chemical Properties

Chemical compositions of metal elements such as Al and Na are measured after digestion of zeolite into the solution by HF. Induced coupled plasma–emission spectroscopy (ICP-ES) is used usually for the analysis of elements. Previously atomic absorption spectroscopy (AAS) was often used especially for alkaline elements, but recently, ICP-ES is preferentially used as a result of improvement of the ICP-ES technique.

Infrared (IR) spectroscopy is measured to see the surface conditions. A stretching vibrations band of Brønsted OH group is observed in the region of 3,600 cm^{-1}, and isolated Si–OH, a fine absorption, is found at 3,745 cm^{-1}. Usually, the region below 1,200 cm^{-1} is less transparent because of an intense absorption of Si–O lattice. IR spectrum after adsorption of pyridine is used to differentiate between Brønsted and Lewis acidity. Although it has been known that the adsorption of pyridine is affected by steric hindrance in the micropores, this method has been used for semiquantitative analysis of Brønsted and Lewis acid sites [6]. Decomposition of adsorbed pyridine molecule is another problem disturbing the quantitative analysis.

Microcalorimetry operated at a high temperature (e.g., 473 K) can determine the amount of adsorbed ammonia molecule and the heat of ammonia adsorption on zeolite [7]. Early studies of the adsorption microcalorimetry were suffered from the low accuracy, and improvements have been attempted. This method needs a long time to wait for a complete equilibrium achievement, and this often causes a practical problem.

2.2 Fundamentals for the TPD of Ammonia

Temperature-programmed desorption (TPD) of ammonia is a method most frequently utilized for the characterization of zeolite acidity. Information on the amount of adsorbed ammonia and the heat of ammonia adsorption can be obtained by the TPD as well as the microcalorimetry but more easily and quickly by the TPD. However, it is not easy to measure the acidity correctly by means of this method. Thus, critical comments have been raised on the method, e.g., as reported by Gorte [8]. In the present chapter, we want to summarize our previous studies on the TPD method as a quantitative tool for the characterization of zeolite acidity [9–11]. Number and strength of the acid sites on zeolites are precisely measured by a usual method of ammonia TPD. Furthermore, this method is effectively advanced to simultaneous measurements of the IR spectroscopy, as mentioned in Chap. 3. Not only ammonia in the gas phase but on the solid is followed during the TPD measurements, from which the structure of acid site is also provided. Fundamentals of the method are studied from experiments and theoretical approaches, on which the practical measurements depend strongly.

2.2.1 Experimental Apparatus of the TPD Method

Figure 2.1 shows an example of the experimental apparatus. An apparatus must be designed to detect desorption profile of ammonia correctly. To install the experimental apparatus, following items are required to be taken care of:

- Metal pipe and connector are to be avoided, because the interaction of ammonia with the metal surface cannot be avoided, even when it is made of a stainless steel. All glass or glass lining pipe installation is recommended. If it is necessary to use metal parts, it is recommended to keep the parts at 373 K in order to suppress the adsorption of ammonia.
- Vacuum pumps are installed before and after the TPD cell; the former is available for pumping the sample before the measurement, and the latter is used to keep the pressure inside the cell less than ambient pressure, for example, 100 Torr (=13.3 kPa). Thus, carrier gas He is passed through the cell in an enhanced rapid flow rate under the evacuated conditions. When the TPD is measured in a slow flow rate, a high temperature is required to desorb ammonia completely, thus leading to the destruction of zeolite structure during the TPD measurement. Experiments of the TPD should be done at the temperatures less than 873 K.
- A liquid nitrogen trap is recommended to install between the vacuum pump and the sample, because the oil mist from the pump may be adsorbed on the sample to disturb the measurements.
- A TPD cell is made of quartz to measure the profile of desorption at temperatures above 773 K. Carrier gas is at first lead into the outside part of the cell to warm up enough, and then to the sample at the inside part of the cell, as illustrated

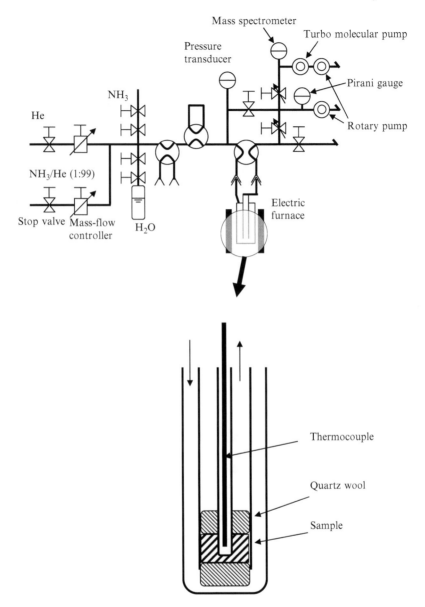

Fig. 2.1 An apparatus of ammonia TPD and the cell with sample

by Cvetanović and Amenomiya [12]. Temperature must be measured with a thermocouple attached to the sample separated through a thin glass wall.
- A mass spectrometer (MS) is recommended as a detector of ammonia, because only ammonia desorption can be measured with MS operating at m/e (mass/charge ratio) = 16. Use of the signal at $m/e = 17$ is not recommended,

2.2 Fundamentals for the TPD of Ammonia

because it is interfered by the fragmentation of water molecule. Both mass numbers of 16 and 17 are usually measured simultaneously, because, though not often needed, it is required to confirm the profile of desorption at m/e of 16 being almost the same as those of 17. When such an oxygen-containing gas as CO_2 is contained in the desorbed material, a fragment with $m/e = 16$ may disturb the analysis. Other detectors, such as thermal conductivity detector (TCD), are not recommended, because it is impossible to identify the desorbed molecule.

As an example, the experimental conditions usually adapted are shown below:
Sample weight is 0.1 g. Powder or granule sample is used.
Flow rate of carrier (usually He) is 60 cm^3 min^{-1} ($= 10^{-6}$ m^3 s^{-1}) under an atmospheric pressure (1.013×10^5 Pa), and pumped into 100 Torr ($= 1.33 \times 10^4$ Pa). Flow rate of the carrier inside the cell is therefore $\dfrac{10^{-6} \times 1.013 \times 10^5}{1.33 \times 10^4}$ m^3 s^{-1} $= 7.6 \times 10^{-6}$ m^3 s^{-1}.

The sample is evacuated at 773 K for 1 h before the measurements. Ammonia is then admitted at 373 K for the adsorption, followed by pumping the gaseous ammonia for ca. 30 min.

Temperature is elevated from 373 K in a ramp rate of 10 K min^{-1} ($= 0.167$ K s^{-1}) until all the ammonia is desorbed completely.

2.2.2 Identification of the Desorption, l- and h-Peaks

Identification of desorption peaks is recently made by the simultaneous measurements of infrared spectroscopy, i.e., IRMS-TPD experiment. However, the explanation of previously studied experiments to identify desorption peaks is helpful to understand the TPD experiment.

In a usual experiment of ammonia TPD on a typical H-form zeolite, such as HZSM-5 and H-mordenite, two desorptions are observed and named *l*- and *h*-peaks for those of ammonia desorbed at low and high temperatures, respectively. Figure 2.2 shows an example of TPD profile observed on H-mordenite, HZSM-5 and H-β. The *l*-peak is, however, not assignable to ammonia adsorbed directly on the acid site, but weakly on the NH_4^+ cation which has been adsorbed on Brønsted acid site. The intensity depends on the experimental conditions of W/F (W, weight of sample, kg; F, flow rate of carrier gas, m^3 s^{-1}), i.e., contact time of the carrier gas, and it becomes small at a small W/F. Strong and effective evacuation removes the *l*-peak completely, but usually the intensity decreases with decreasing the W/F. At a small weight of zeolite and/or a fast flow rate of carrier, the degree of evacuation becomes high to remove fully the ammonia weakly adsorbed on the zeolite. Physical adsorption of NH_3 or $(NH_3)_n$ polymeric species on NH_4^+ was revealed by an NMR study [13]. Therefore, the intensity of the *l*-peak cannot be measured quantitatively. In other words, the intensity of the *l*-peak does not provide us with any physical meaning at all.

Usually, the *h*-peak shows the desorption profile of ammonia which had been adsorbed on the acid sites. The intensity of *h*-peak does not depend on the experimental

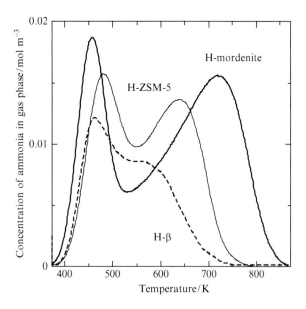

Fig. 2.2 Examples of TPD profile observed on H-mordenite, H-ZSM-5, and H-β

conditions, therefore, we can measure the amount of ammonia desorbed in a unit of mol kg^{-1} of zeolite, quantitatively.

On the other hand, a large l-peak appears on such a cation exchanged zeolite as NaH-zeolites, because NH$_3$ is adsorbed on Na cation weakly. Na$^+$ cation could be regarded as a Lewis acid site with a property of the electron acceptor. However, the quantitative measurements have not been studied sufficiently.

2.2.3 Identification of Ammonia Desorbed from Y Zeolite

It is easy to identify desorption peaks from the zeolite having the strong acidity, because the h-peak could be distinctively identified. On the Y zeolite, however, it is difficult to identify ammonia desorbed from the acid site. Because the Y zeolite has the acidity so weak that the desorbed ammonia is observed at low temperatures, thus overlapping with the l-peak. In other words, it is impossible to discriminate between the portions of ammonia adsorbed on acid sites and on NH$_4$$^+$ cation because of the fully overlapped profile. To remove only the portion of ammonia on the NH$_4$$^+$, a method of water vapor treatment has been conducted. Principle of the method is that water is adsorbed on the NH$_4$$^+$ in preference to NH$_3$ [14]. Therefore, an input of humidity into He carrier gas during the TPD measurement or a treatment immediately after the adsorption of ammonia is performed to replace the removable ammonia with water, thus remaining only the h-peak assignable to the acid site [15]. Figure 2.3 shows an example of ammonia TPD on H-mordenite without and with a water vapor treatment. Two desorptions are distinctively observed as l- and h-peaks

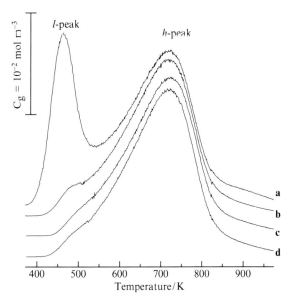

Fig. 2.3 Ammonia TPD on H-mordenite without (**a**) and with a water vapor treatment, one, two, and three times conducted (**b**), (**c**), and (**d**), respectively

at low and high temperatures, respectively. The intensity of the *l*-peak disappears gradually by conducting the treatment by water vapor, while that of the *h*-peak does not change. It is therefore confirmed that only the *l*-peak is removed by the water vapor treatment. Then, this method is applied to the H-Y zeolite in Fig. 2.4. By the conventional method shown in Fig. 2.4a, an ammonia desorption peak is observed at ca. 450 K, and it is impossible to discriminate between *l*- and *h*-peaks. With the water vapor treatment, the intensity of desorbed ammonia decreases, because the portion of *l*-peak is removed. Therefore, we identify the profiles of acidity on the H-Y zeolite from the TPD of ammonia observed after the water vapor treatment. Bagnasco also reports a similar method to remove the *l*-peak selectively using the water vapor treatment [16].

2.3 Theory for the TPD of Ammonia

2.3.1 Conditions of the Equilibrium

Cvetanović and Amenomiya classified the TPD experiments into three categories [12], in which the TPD experiments are controlled by

1. Kinetics of desorption
2. Equilibrium between molecules in the gas phase and on the solid or
3. Diffusion of molecule

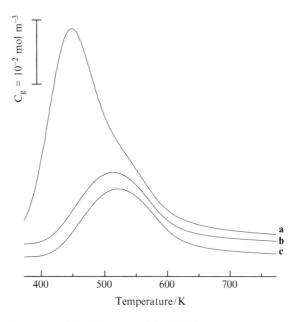

Fig. 2.4 Ammonia TPD on H-Y zeolite without (**a**) and with a water vapor treatment once (**b**) and twice (**c**) conducted

Among three categories, usual TPD experiments on zeolites are measured at the conditions of 2: equilibrium in which ammonia readsorption occurs freely. Figure 2.5 shows an experimental evidence to show the conditions of ammonia TPD. Based on a research project planned and organized by the Committee of Reference Catalyst, Catalysis Society of Japan, ammonia TPD was studied among people belonging not only to the universities but also to the companies. People belonging to more than ten groups measured the TPD of ammonia on a zeolite which was distributed as a Reference Catalyst, and reported their measurements to the Committee. It was thus found that the temperature of ammonia desorption from H-mordenite depended on a parameter of W/F; i.e., the larger the W/F, the higher the peak temperature. The difference in the reported temperatures is more than 200 K by varying W/F from 10^{-5} to 10^{-2} g min cm^{-3}. The strong dependence of the peak temperature upon the W/F suggests that the TPD experiment is controlled by the equilibrium, as mentioned precisely below. It is sure to indicate that the peak temperature of desorption is not a parameter specific for the measured zeolite. In other words, the strength of the acid site cannot be measured from the desorption temperature directly. When the samples are measured under the common experimental conditions, the strength of acid site is not measured from the peak temperature precisely; because it actually depends on a parameter of $A_0 W/F$ (A_0, number of acid sites in mol kg^{-1}), as explained below. The temperature for the TPD experiment shifts high with increasing number of acid sites and/or the contact time of carrier gas, as mentioned below. This is an important conclusion for ammonia TPD as a method of characterization of zeolite acidity.

2.3 Theory for the TPD of Ammonia

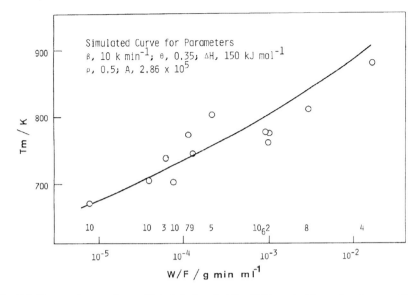

Fig. 2.5 Temperature maximum (T_m) plotted against the W/F, contact time of carrier gas with the zeolite sample based on the study by various research groups shown by the numbers 2–10. A curve for the relation is simulated based on the equation proposed by Cvetanović and Amenomiya [12],
$$2\log T_m - \log \beta = \frac{\Delta H}{2.303 R T_m} + \log Q, \quad Q = \frac{(1-\theta)^2 V \Delta H}{FAR} \text{ and } V = \frac{W}{\rho}, \text{ using parameters,}$$
ramp rate of temperature (β), coverage by ammonia at the T_m (θ), density of zeolite (ρ), and an assumed parameter A

2.3.2 Derivation of a Fundamental Equation

Thermodynamics and kinetics equations shown below are summarized into a fundamental equation of ammonia TPD. Because the equilibrium is attained between ammonia molecules on the surface and in the gas phase during the experiment, (2.1) is written as follows:

$$(NH_3)_a \rightleftarrows NH_3(g) + (\ )_a \quad (2.1)$$

where $(NH_3)_a$ and $(\ )_a$ show ammonia adsorbed on the site and vacant site, respectively. An equilibrium constant K can be written as

$$K = \frac{1-\theta}{\theta} \frac{P_g}{P^0} = \frac{1-\theta}{\theta} \frac{RT}{P^0} C_g \quad (2.2)$$

where θ is coverage by ammonia on the acid sites, P_g and C_g are partial pressure (Pa) and concentration of ammonia in the gas phase (mol m^{-3}), respectively, and P^0, R, and T are standard pressure (1.013×10^5 Pa), gas constant (8.314 J K^{-1} mol^{-1}), and temperature (K), respectively. On the other hand, material balance of ammonia molecule is kept always in the measurement system. In other words, sum of the changes of ammonia molecules in the gas phase and on the surface during the experiment must be zero, and therefore

$$FC_g = -A_0 W \frac{d\theta}{dt}. \qquad (2.3)$$

From (2.2) and (2.3), it follows that

$$C_g = -\frac{A_0 W}{F} \frac{d\theta}{dt} = \frac{\theta}{1-\theta} \frac{P^0}{RT} K_p. \qquad (2.4)$$

Because temperature is elevated in a ramp rate $\beta (= dT/dt,\ \text{K s}^{-1})$, and also from the Gibbs free energy relation, we can derive

$$C_g = -\frac{\beta A_0}{F} \frac{d\theta}{dT} = \frac{\theta}{1-\theta} \frac{P^0}{RT} \exp\left(-\frac{\Delta H}{RT}\right) \exp\left(\frac{\Delta S}{R}\right), \qquad (2.5)$$

where ΔH and ΔS are changes of enthalpy and entropy upon desorption of ammonia, respectively. Equation (2.5) shows a change of ammonia concentration in the gas phase with respect to the temperature, i.e., it is exactly the same as the TPD profile. A fundamental equation of ammonia TPD is thus derived.

A simulated spectrum can be obtained from (2.5) based on the assumption of a set of appropriate parameters, and it is benefit to explain the characteristics of the TPD spectrum. Figure 2.6 shows simulated spectra of ammonia TPD which are obtained at different numbers of acid site, A_0. Following important remarks are noticeable from the simulation.

Temperature of the desorption shifts high with increasing A_0, when a constant ΔH is assumed. This means that the temperature of desorption does not correspond to the strength of acid site, as mentioned above. TPD profile with a constant ΔH is

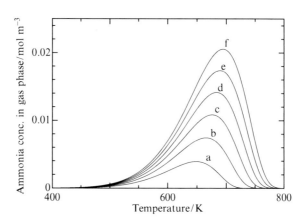

Fig. 2.6 Simulated TPD spectra with assumed parameters; W, 1×10^{-4} kg; F, 7.6×10^{-6} m^3 s^{-1}; ΔH, 140 kJ mol^{-1}; ΔS, 150 J K^{-1} mol^{-1} and variable parameters of A_0 of 0.2 (**a**), 0.4 (**b**), 0.6 (**c**), 0.8 (**d**), 1.0 (**e**), and 1.2 (**f**) mol kg^{-1}

2.3 Theory for the TPD of Ammonia

similar to the experimental observation. Therefore, the desorption profile with about 150 K of half width temperature, as shown in Fig. 2.6, is not caused by the distribution of acid strength, but by the readsorption of ammonia. Actually, a small standard deviation of ΔH is required to do the curve fitting calculation, because the simulated spectrum has the width which is sharper than experimentally observed (*vide infra*).

2.3.3 Determination of ΔH and Constancy of ΔS (Desorption)

At the temperature of desorption peak, $dC_g/dT = 0$, therefore the differentiation of C_g shown by (2.5) with respect to T gives us

$$\ln T_m - \ln \frac{A_0 W}{F} = \frac{\Delta H}{RT_m} + \ln \frac{\beta(1-\theta_m)^2(\Delta H - RT_m)}{P^0 \exp(\Delta S/R)}, \quad (2.6)$$

where T_m and θ_m are each parameters at the peak maximum. The logarithm term in the right side of (2.6) changes little by varying values of the parameter W/F, because $RT_m \ll \Delta H$, θ_m changes small, and the logarithm of these variable parameters is calculated. Therefore, it may be regarded as a constant. Thus, experiments by varying the W/F are conducted; and plotting the value of the left side of (2.6) against $1/T_m$ gives a straight line with a slope of $\Delta H/R$, from which the ΔH is determined. This method of determination of ΔH was applied to H-mordenite and H-ZSM-5, as shown in Fig. 2.7. Measured values of ΔH are satisfactorily consistent with the values on the same zeolite measured by a microcalorimeter [17]. Thus, the method stated above is supported by another experiment.

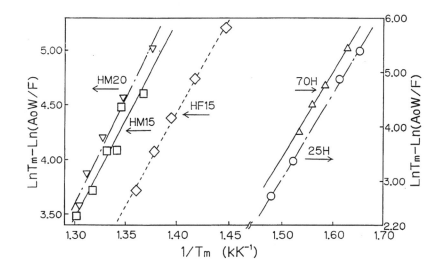

Fig. 2.7 Determination of ΔH applied to mordenite (HM*n*), ferrierite (HF*n*), and ZSM-5 (*n*H), where numbers show the silica to alumina molar ratio of the zeolites

Once when the ΔH is determined, ΔS is calculated from (2.6). ΔS thus measured is, however, almost constant independent of zeolite species. The interesting finding of ΔS constancy suggests the physical chemistry of desorption of ammonia from the solid. The entropy increases not only upon the desorption of ammonia, but also upon the mixing with the He carrier, i.e.,

$$\Delta S = \Delta S(\text{desorption}) + \Delta S(\text{mixing}), \tag{2.7}$$

where $\Delta S(\text{mixing})$ for ammonia molecule is calculated from the concentration of ammonia in the gas phase given by (2.8):

$$\Delta S(\text{mixing}) = -R \left[\ln x_{NH_3} + \frac{x_{He}}{x_{NH_3}} \ln x_{He} \right], \tag{2.8}$$

where x denotes the mole fraction of ammonia or He. Be aware that $\Delta S(\text{mixing})$ thus calculated is the change of entropy of ammonia in the mixing procedure. $\Delta S(\text{desorption})$ is thus calculated to be ca. 95 J K^{-1} mol^{-1}. This value of $\Delta S(\text{desorption})$ is very close to ΔS for vaporization of liquid NH$_3$, 97.2 J K^{-1} mol^{-1}. Similar values of ΔS suggest that liquid and adsorbed phases of ammonia are in similar entropy levels. Phase transformations, vaporization and desorption, both accompany an increase of translational entropy primarily, but changes of rotation and vibration entropies are so small that these could be disregarded. Therefore, the physical chemical consideration accounts for $\Delta S(\text{desorption})$ constancy of ca. 95 J K^{-1} mol^{-1}, and strongly supports the validity of the method. The $\Delta S(\text{desorption})$ constancy upon desorption is the most fundamental and important principle for the ammonia TPD to measure the zeolite acidity quantitatively [18].

2.4 Practical Measurements of Ammonia TPD

2.4.1 Curve Fitting Method to Measure the Strength of Acid Site

Simulation of ammonia TPD based on (2.5) can be performed on a personal computer. A spreadsheet program such as Microsoft Excel can be used for the simulation. Coverage θ decreases from 1 to 0, as the temperature is elevated during the TPD experiment; but actually the initial value of θ is set to be a slightly smaller value such as 0.999 at the start-up temperature, e.g., 373 K, because the term $(1 - \theta)$ in a denominator must be positive. Because digital parameters of experimental observation, time, temperature, and intensity of MS, are measured and transferred to a computer at every finite interval, the calculation is performed at every step to compare with the observed spectra. The change of coverage by ammonia θ with increasing the temperature can be described as

$$\theta_{i+1} = \theta_i + \left(\frac{d\theta}{dT} \right)_i \Delta T. \tag{2.9}$$

2.4 Practical Measurements of Ammonia TPD

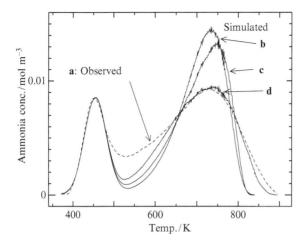

Fig. 2.8 Observed (**a**) and simulated spectra from the assumption of constant ΔH and ΔS (**b**) and of constant ΔH and corrected ΔS (**c**), and of small-distributed $\Delta H(142 \pm 6\,\text{kJ mol}^{-1})$ and corrected ΔS (**d**)

$d\theta/dT$ is calculated from (2.5), and, in addition, ΔS is given from (2.7) and (2.8). Therefore, the simulated spectrum of ammonia desorption is obtained with assumed parameters of A_0 and ΔH. To compare with the experimental TPD spectrum, the value of A_0 is known from the h-peak and only the ΔH is an unknown parameter.

Figure 2.8 shows a comparison of thus simulated spectrum with the experimentally observed one on the H-mordenite. The simulated spectrum shows a sharp TPD profile against the temperature compared with the actual observation. This is because, obviously, a constant ΔH only is provided, and any distribution of the acid strength is not taken into consideration. Therefore, the Gaussian distribution with a standard deviation is produced, and sum of the distributed intensities is compared with the experimental spectrum. Seven fractions of the Gaussian distribution are produced with $\Delta H + n\sigma (n = -3, -2, -1, 0, 1, 2, 3; \sigma$, standard deviation), and sum of the simulated spectrum is compared with the experimental observation. The simulated spectrum with the ΔH of $142 \pm 6\,\text{kJ mol}^{-1}$ is shown in Fig. 2.8d; and thus simulated profile is fitted well to the experimental observation (Fig. 2.8a). Thus, ΔH as an index of acid strength is determined as above. The small value of the standard deviation is noteworthy, meaning the small distribution of the acid strength in the zeolite. Fine distribution of acid strength is an important profile in a usual zeolite, which differs greatly from those of metal oxides.

2.4.2 ΔH on Various Zeolites with Different Concentrations of Acid Site

The acid strengths of H-mordenite, H-ZSM-5, H-β, H-Y, and MFI-gallosilicate with different numbers of acid sites are measured by the method of curve fitting

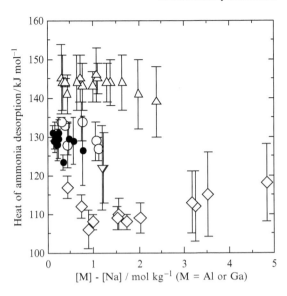

Fig. 2.9 Dependence of heat of ammonia desorption (ΔH) against the composition over mordenite (*open triangle*), ZSM-5 (*open circle*), β (*open inverted triangle*), Y (*open diamond*), and MFI-gallosilicate (*closed circle*). The vertical bar shows the standard deviation of ΔH

mentioned above. Zeolite samples with different concentrations of acid site are prepared by the synthesis with varying the Si/Al ratio in the parent gel, the ion exchange of Na cation with proton, or the dealumination using HCl. Thus measured ΔH is summarized in Fig. 2.9. From the measurements, it is found that the strength of acid site does not change largely with changing the number of acid sites, but it depends on the zeolite species [18]. The ΔH on HZSM-5 is nearly constant, 130 ± 8 kJ mol^{-1}, independent of the concentration of acid site. Those on H-mordenite are, however, decreased small in the high concentration of acid sites. Smaller values of ΔH observed on the large number of acid sites on H-mordenite are difficult to explain only based on usual measurements of ammonia TPD. This method is applied to β-zeolites synthesized by a dry-gel conversion method, in which the number of acid sites is varied from 0.20 to 0.80 mol kg^{-1} [19]. The values of ΔH on the β zeolites are almost constant, 124–127 kJ mol^{-1}, similar to on HZSM-5 and H-mordenite. Y zeolites prepared by changing the acid site concentration are also studied from the same viewpoints. The conclusion for the strength of acid site that we have reached is that the ΔH is independent of the concentration, and shows the almost constant value, ca. 110 kJ mol^{-1} [14]. The strength of MFI-gallosilicate is similar or slightly weaker than the aluminosilicate analogue (HZSM-5) [20].

Principles for the acid strength observed in these studies are summarized below:

- The acid strength is mainly controlled by the crystal type as MOR > MFI > BEA > FAU.
- Small variations of averaged ΔH in one kind of zeolite with varying Al and Na contents are observed, but they are less than 10 kJ mol^{-1} and obviously smaller than differences in those of zeolite species. This indicates that the composition (Al and Na concentrations) does not give a remarkable influence on the acid strength, at least in the experimental range.

2.4 Practical Measurements of Ammonia TPD

The conclusion might be surprising for some researchers, because previously it has been believed that the higher Al concentration gives the weaker acidity [21]. Nevertheless, this is a misunderstanding due to incorrect measurements of acid strength. The misunderstandings also are induced from the following shortages of observations:

- It is sure that the zeolite species such as MFI and MOR, which tends to have highly siliceous composition, has the stronger acidity. This does not tell us that the Al content directly influences the acid strength.
- At a very high Al content, the acid site is destroyed by contact with water vapor in atmosphere as shown in Sect. 2.5.2, and it looks that the acidity is absent on the highly aluminated zeolite. However, as long as the structure of acid site is kept, it shows the acidity, as stated below in this chapter.
- Generation of extra-framework Al with dealumination often enhances the acid strength of framework acid site, as demonstrated on ultrastable Y (USY) in Chap. 3. It looks as if the decrease in framework Al concentration increases the acid strength.
- Small difference in the acid strengths is seen; for example, in MOR, the acid strength decreases with increasing the [Al]–[Na] from 1.5 to 2.5 mol kg^{-1}, as shown in Fig. 2.9.

The correct measurements of acid strength by ammonia TPD methods deny these considerations, and it is concluded that the strength of framework acid site in zeolite is mainly controlled by the crystal type.

Table 2.1 shows the averaged acid strengths of various zeolites. In some cases, the different OH groups are individually measured by means of the IRMS-TPD method [22] as described in detail in Chap. 3. When the distribution of Brønsted acid sites

Table 2.1 Acid strengths of framework OH groups on zeolites with various crystal types

Framework type code and position	ΔH (kJ mol^{-1})
MOR (12-ring)	142
MOR (8-ring)	153
MWW	140
MFI	137
FER (10-ring)	142
FER (6/8-ring)	141
CHA (O2H)	139
CHA (O1H)	136
CHA (O3H)	133
CHA (O1H)	105
*BEA	128
FAU (O1H)	108
FAU (O1'H)	110
FAU (O2H)	119
FAU (O3H)	105

is clearly identified, the dependence of the strength of acidity upon the structure is concluded in more detail, e.g., MOR 8-ring (8-oxygen membered ring) > MOR 12-ring and FAU O2H (sodalite cage) > FAU O1'H ≈ FAU O1H (supercage). The general trend of the structure dependence will be discussed in Chap. 4.

2.5 Number of Acid Sites on Various Zeolites

2.5.1 Number of Acid Sites Correlated with Al Concentration

The number of acid sites is a practical parameter more than the strength of acid site, and it is relatively easy to understand, because the Brønsted acid site is generated due to the mechanism of T-site replacement by Al (isomorphous replacement), as shown by the scheme in Fig. 2.10. Negative charge formed due to the replaced Al must be balanced with such a cation as H^+, thus producing the Brønsted acid site.

Figure 2.11 shows an experimental observation about the relationship between the number of acid sites and Al concentration in the framework of the HZSM-5 and H-mordenite. A clear coincidence is found between them to support the generation mechanism of the acid site due to the Al in the framework, as expected. When Na cations remain, Al–Na is used as a parameter, because Na cation is exchanged with H^+. However, the principle of acid site generation correlated with Al is found only in less than ca. 1.5 mol kg^{-1} of the acid site, and the number of acid sites decreases gradually in more than 1.5 mol kg^{-1} of Al concentration. Siliceous zeolites such as H-ZSM-5 and H-β has the concentration of Al of less than 1.5 mol kg^{-1}; and therefore, the maximum number of acid sites is not found at all. The maximum number of acid sites is observed on mordenite in Fig. 2.11, and particularly remarkable on Y zeolite.

It has been known that the number of acid sites on the Y zeolite shows the maximum value against the concentration of framework Al. Figure 2.12 shows the typical observation of the number of acid sites on the Y zeolite dependent on the concentration of Al. On the left side of the maximum, the number of acid sites increases with the Al concentration. However, the decrease in the number of acid sites with the Al concentration at the right side of the maximum is unexpected from the simple generation mechanism of acid site. The following experiment helps us to understand the peculiar and abnormal behavior of the acid site.

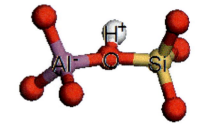

Fig. 2.10 Brønsted acid sites generated by the isomorphous replacement

2.5 Number of Acid Sites on Various Zeolites

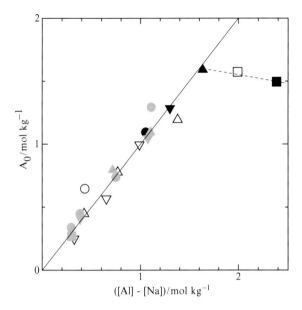

Fig. 2.11 Relationship between number of acid site A_0 and [Al]–[Na] concentration in the framework of HNa-MOR with $Si/Al_2 = 10$ (*dark filled square*), 15 (*dark filled triangle*), and 20 (*dark filled inverted triangle*), dealuminated from H-MOR with $Si/Al_2 = 10$ (*gray filled square*), 15 (*gray filled triangle*), and 20 (*gray filled inverted triangle*), HNa-MFI (*open circle*) and H-MFI (*dark closed circle*) with $Si/Al_2 = 23.8$ and H-MFI samples with $Si/Al_2 > 25$ (*gray closed circle*)

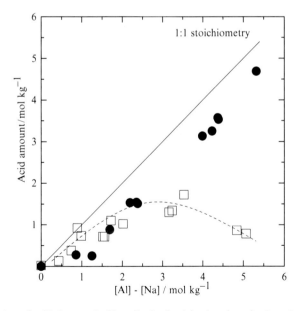

Fig. 2.12 Number of acid sites on the Y zeolite in situ (*close*) and ex situ (*open*) prepared correlated with the concentration of Al–Na

In a usual experiment, the NH$_4$-type zeolite is thermally treated to remove NH$_3$ at a high temperature such as 773 K and to convert into the H-type zeolite; the produced H-type zeolite is usually stored in a bottle at room temperature. The sample stored is then put into a measurement cell or a reactor for the run of measurement or reaction, respectively. The measurement data shown in Fig. 2.12 with the maximum number of acid sites against the Al concentration are obtained at the usual procedure of the TPD measurements. However, the H-type zeolite prepared in situ in the TPD cell and measured without exposing the atmosphere shows the value completely different from those of above mentioned. Those in situ prepared samples show the number of acid sites almost proportional to the Al concentration in the framework; in other words, an abnormal decrease of the acid site with Al is not found at all. This finding indicates clearly that the number of acid sites in a large concentration of Al (>1.5 mol kg^{-1}) depends on the method of preparation [23].

2.5.2 In Situ and Ex Situ Prepared H-type Zeolites

In situ prepared samples that we named are those of H-type zeolite prepared in situ in the reactor and not exposed to atmosphere. In contrast, the usual H-type zeolites are named ex situ prepared one, because the H-type zeolite is prepared before the measurement ex situ in the reactor. The difference in the preparation procedures, in situ and ex situ preparations, strongly affects the number of acid sites only at the large concentration of framework Al.

The reason for the difference is studied, and it is found that the moisture contained in the atmosphere results in dealumination and a decrease in the number of acid sites [24]. Even a small concentration of water causes the destruction of zeolite structure. Influence of the humidity upon the zeolite structure is not expected quantitatively, but it is sure that the degree becomes large at the high concentration of Al; and this may be related to the hydrophobic property due to the large concentration of Al.

In situ preparation of H-type zeolite, on the other hand, exhibits the number of acid sites which is nearly equal to Al even at its large concentration. Therefore, the Al atom included in the framework contributes to the formation of acid site, and the number of acid sites increases to ca. 5 mol kg^{-1}. In situ prepared H-type zeolite preserves the solid acidity dignity, as long as it is kept unexposed to the moisture. However, the high concentration of the solid acidity is not available under the usual conditions, because so usually the zeolite sample is used in the presence of humidity. Therefore, the large number of the acid sites is not available in use practically. The principle of the formation of acid site in the in situ H-type zeolite is important to understand the generation mechanism of the Brønsted acid site.

The in situ preparation of H-type zeolite is important, when it is utilized for the IR measurements. The IR spectra on the in situ H-type zeolite are fine and clean; therefore, it is advantageous to study the IR spectra precisely. This is true also for the zeolite with a low concentration of Al. Therefore, it is expected that a subtle change of surface structure occurs under the presence of water.

An interesting subject that we learn from the maximum number of acid sites is the surface density of acid site. Because the zeolite has the surface area of ca. 400 m^2 g^{-1}, the acid site concentration, 1.5 mol kg^{-1}, corresponds to ca. 2 nm^{-2} of the surface concentration. The maximum surface concentration found in the zeolite is almost the same as the maximum surface concentration observed on the metal oxides. Studies on the acid site generation on the metal oxide monolayer show that the maximum surface concentration is observed when the support surface is fully covered by the monolayer with the surface concentration, ca. 2 nm^{-2} [25]. The coincidence between them may indicate how the acid site is maintained in the presence of water vapor. It is expected that the acid sites would be collapsed easily at the conditions of more than 2 nm^{-2} of the surface acid site concentration, maybe due to the mutual interaction of OH groups.

References

1. G.T. Kerr, Zeolites **9**, 350 (1989)
2. F. Gramm, C. Baerlocher, L.B. McCusker, S.J. Warrender, P.A. Wright, B. Han, S.B. Hong, Z. Liu, T. Ohsuna, O. Terasaki, Nature **444**, 79 (2006)
3. J. Klinowski, J.M. Thomas, C.A. Fyfe, G.C. Gobbi, Nature **296**, 533 (1982)
4. D. Freude, H. Hunger, H. Pfeifer, W. Schwieger, Chem. Phys. Lett. **128**, 62 (1986)
5. M. Inomata, M. Yamada, S. Okada, M. Niwa, Y. Murakami, J. Catal. **100**, 264 (1986)
6. T.R. Hughes, H.M. White, J. Phys. Chem. **71**, 2192 (1967)
7. Y. Mitani, K. Tsutsumi, H. Takahashi, Bull. Chem. Soc. Jpn. **56**, 1917 (1983)
8. R.J. Gorte, Catal. Lett. **62**, 1 (1999)
9. M. Niwa, N. Katada, Catal. Surv. Jpn. **1**, 215 (1997)
10. N. Katada, M. Niwa, Catal. Surveys Asia **8**, 161 (2004)
11. M. Niwa, N. Katada, J. Jpn. Petrol. Inst. **52**, 172 (2009)
12. R.J. Cvetanović, Y. Amenomiya, Adv. Catal. **17**, 103 (1967)
13. W.L. Earl, P.O. Fritz, A.A.V. Gibson, J.H. Lunsford, J. Phys. Chem. **91**, 2091 (1987)
14. H. Igi, N. Katada, M. Niwa, in *Proceedings of the 12th International Zeolite Conference*, vol. 4, ed. by M.M.J. Treacy, B.K. Marcus, M.E. Bisher, J.B. Higgins (Materials Research Society, Warrendale, 1999), p. 2643
15. A.W. Chester, J.B. Higgins, G.H. Kuehl, J.L. Schlenker, in *Preprint of 11th Discussion on Reference Catalyst* (Catalysis Society of Japan, Tokyo, 1987), p. 22
16. G. Bagnasco, J. Catal. **159**, 249 (1996)
17. M. Niwa, N. Katada, M. Sawa, Y. Murakami, J. Phys. Chem. **99**, 8812 (1995)
18. N. Katada, H. Igi, J.H. Kim, M. Niwa, J. Phys. Chem. B **101**, 5969 (1997)
19. Y. Miyamoto, N. Katada, M. Niwa, Microporous Mesoporous Mater. **40**, 271 (2000)
20. T. Miyamoto, N. Katada, J.H. Kim, M. Niwa, J. Phys. Chem. B **102**, 6738 (1998)
21. P.A. Jacobs, Catal. Rev. Sci. Eng. **24**, 415 (1982)
22. K. Suzuki, T. Noda, N. Katada, M. Niwa, J. Catal. **250**, 151 (2007)
23. N. Katada, Y. Kageyama, M. Niwa, J. Phys. Chem. B **104**, 7561 (2000)
24. N. Katada, T. Kanai, M. Niwa, Microporous Mesoporous Mater. **75**, 61 (2004)
25. M. Niwa, Y. Habuta, K. Okumura, N. Katada, Catal. Today **87**, 213 (2003)

Chapter 3
IRMS-TPD Measurements of Acid Sites

Abstract An improved method of ammonia TPD, named IRMS-TPD, i.e., simultaneous measurements of ammonia probe adsorbed on the zeolite and desorbed in the gas phase by means of infrared and mass spectroscopies, respectively, is elucidated. Distribution of Brønsted acid sites on pure and modified zeolites is measured precisely and its dependence on the zeolite structure is clarified.

3.1 Measurement Method

3.1.1 What Is Obtained from the TPD Measurement

As mentioned in Chap. 2, ammonia TPD measurements provide us with information about number and strength of acid site. With careful experiments conducted, precise and useful information is provided. However, a serious drawback exists in the TPD measurements because no information on the acid site structure is provided at all. Therefore, another experiment has to be conducted together with the TPD experiments. In the TPD measurements, a differentiation between Brønsted and Lewis acid sites is impossible. Thereby, such an additional experiment as infrared observation of the adsorbed pyridine must be conducted to differentiate between Brønsted and Lewis acid sites. Direct and simultaneous measurements, if possible, would be carried out to characterize the zeolite acidity in more detail. When information of the structure of acid site is added to the ammonia TPD measurements, a significant progress in the characterization is anticipated. In this chapter, infrared/mass spectroscopy-ammonia TPD (IRMS-TPD) measurement, which is a method of total characterization, is described precisely.

3.1.2 Experimental Methods

An idea to develop the IRMS-TPD experiments is simple, but challenging; i.e., an infrared spectroscopy is simultaneously measured to follow the change of adsorbed ammonia species during ammonia TPD. Infrared spectroscopy is a method to provide us principally with qualitative information of the studied molecule and material, but the precisely quantitative measurements are not easy to do. On the other hand, mass spectroscopy is an analytical instrument which enables us to measure the gaseous component quantitatively. Therefore, infrared and mass spectroscopies are complementary, when both instruments are working simultaneously. IRMS-TPD experiments are therefore designed so adequately to complement the drawback of each methodology.

Infrared spectroscopy is measured through a thin wafer of the sample; therefore, high transparency of infrared beam is indispensable. Because the sample weight is so small and only 5–20 mg, mass spectroscopy must keep a high sensitivity to enable us to detect a small change of the intensity. These difficult requirements are however overcome. Current progresses of IR and MS spectroscopies allow us to measure both simultaneously.

Figure 3.1 shows the experimental apparatus of the ammonia IRMS-TPD. Because IR and MS spectroscopies must be measured at the same time, the so-called dead volume of the instrument must be as small as possible, and the carrier gas helium is flown at a fast flow rate. The time lag of IR measurements becomes shortened. Thus, the flow rate of the carrier is ca. ten times faster than usual, and the amount of sample is ca. one-tenth; thus, the experimental conditions of W/F are about one hundredth of the usual (MS-) TPD measurement shown in Chap. 2.

Before the ammonia adsorption, reference spectra are measured at every 10 K with raising the temperature up to the maximum in a ramp rate of $10\,\text{K}\,\text{min}^{-1}$. Because some of IR profiles change with the temperature, the measurements of reference spectra without adsorbed ammonia are necessary. Ammonia is then adsorbed on the evacuated sample, usually at 373 K, followed by evacuation. Temperature is again elevated; infrared spectra are measured at every 10 K, and mass spectroscopy is working to measure the change of ammonia concentration in the gas phase. Reference spectra, $A(T) - N(T)$, are calculated after the experiments, where $A(T)$ and $N(T)$ are IR absorbances at a temperature T before and after ammonia adsorption,

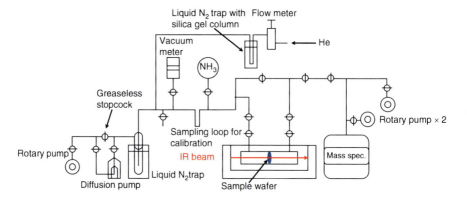

Fig. 3.1 Experimental apparatus of the ammonia IRMS-TPD

respectively. At the selected band positions, differential changes of the reference spectra against the temperature, i.e., $d\{A(T) - N(T)\}/dT$, are calculated, which we call the IR-TPD of adsorbed ammonia or OH bands. The IR-TPD is compared with the MS-TPD, and on the basis of the comparison, the number and strength of individual acid sites are measured. Practical measurement procedures are described in detail as follows.

3.1.3 Required Corrections, IR Band Position and ΔS

Two important corrections are required in the IRMS-TPD experiment [1].

The OH band position shifts to low wavenumber by increasing the temperature. It is found experimentally that the band position of OH shifts almost linearly with the measurement temperature. Therefore, a linear equation is used to correct the band position, i.e.,

$$\Delta v/\Delta T = -a(\text{cm}^{-1}\,\text{K}^{-1}), \tag{3.1}$$

where Δv and ΔT are the changes of band position and temperature, and a is a parameter, equal to 0.034(H-Y) – 0.056(H-mordenite).

ΔS constancy is a principle for the ammonia TPD under the equilibrium conditions to measure the ΔH, as mentioned in Chap. 2. However, the equilibrium between ammonia molecules on the sample and in the gas phase is not fully satisfied at $W/F < 0.5\,\text{kg s m}^{-3}$, where usually the IRMS-TPD is operated. Figure 3.2b shows the plot of confirmation of equilibrium in the TPD experiment, and the equilibrium is satisfied under the conditions shown by the linear plot. As shown by the figure, a deviation from the linear relation is observed under the fast flow rate, where

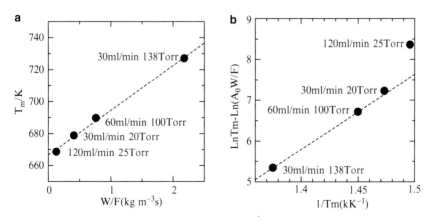

Fig. 3.2 (a) Dependence of the temperature maximum (T_m) upon the conditions of W/F; (b) derived plot of equilibrium confirmation in the TPD experiment, where the equilibrium is satisfied under the conditions shown by the linear plot

the IRMS-TPD experiment is conducted actually. Because of the requirement for a fast flow rate and a small weight of sample, the conditions of equilibrium between ammonia molecules in the gas phase and on the surface are not satisfied sufficiently. ΔS is therefore corrected so as to determine the ΔH measured in the IRMS-TPD which is exactly equal to the ΔH in usual ammonia TPD. ΔS value correction will be described in the next chapter exemplified with a study on H-mordenite. The ΔS correction will not be required, however, when the experimental apparatus is designed with a time gap between IR and MS made small, because at the conditions the flow rate of carrier gas can be decreased.

3.2 Proton Form Zeolite

3.2.1 H-Mordenite [1]

Mordenite has a large amount of strong acid sites, and it is easy to measure the solid acidity. Experimental observations and following procedures for measurements of IRMS-TPD are, therefore, explained using an example of H-mordenite. Figure 3.3 shows the difference IR spectra obtained at 373–773 K during the TPD experiments. A fine absorption at ca. 1,450 cm^{-1} is ascribable to the bending vibration of NH_4^+, and a broad absorption from 3,500 to 1,500 cm^{-1} is due to the stretching vibration of N–H bond which is combined with the vibration of Si–O in the zeolite lattice. In the OH region, a negative absorption band profile is observed, because the OH intensity diminishes completely by the adsorption of ammonia. It is easy to identify that the OH band lost the intensity by accommodating NH_3 to become NH_4^+. With elevating the temperature, the NH_4^+ intensity decreases and the OH intensity recovers because of the desorption of ammonia as shown in the equation: $(O^-NH_4^+) \rightarrow NH_3 + (O^-H^+)$.

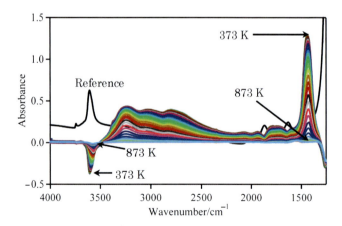

Fig. 3.3 Difference IR spectra obtained at 373–873 K during the TPD experiments on *in situ* prepared HM-15 (H-mordenite with 15 of Si/Al$_2$ ratio)

3.2 Proton Form Zeolite

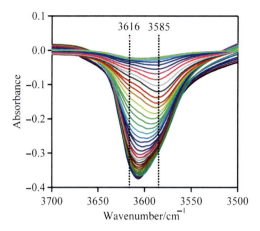

Fig. 3.4 Enlarged portion of the OH band intensity in the difference spectra

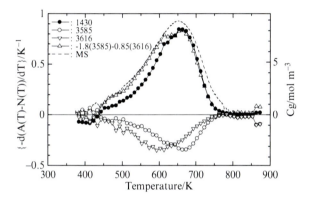

Fig. 3.5 Comparison of the calculated IR-TPD with MS-TPD (C_g)

As an enlarged portion of the OH band intensity is shown in Fig. 3.4, the band position shifts to low wavenumber with increasing the temperature. The band position of this figure has been corrected based on the equation shown above. The shift of band position of OH suggests that two kinds of OH change their intensities with changing the temperature and the sites' distribution depends on the temperature. It is identified that two kinds of OH are observed at 3,616 and 3,585 cm^{-1}. Therefore, the OH band intensity is divided into two portions at these wavenumbers with Gaussian distributions assumed. Figure 3.5 shows a comparison of thus calculated IR-TPD with MS-TPD. First, a coincidence of IR-TPD of NH_4^+ with MS-TPD is noteworthy; and this clear finding suggests that the studied mordenite has the Brønsted acidity preferentially. The IR-TPD of NH_4^+ should have a mirror image in relation with IR-TPD of two OH bands because the Brønsted acid sites are formed on the hydroxides. Because the hydroxide bands have different extinction coefficients, parameters to cancel the difference in extinction coefficients are required. It has been previously reported that the H-mordenite has two kinds of Brønsted OH located at 8- and 12-rings, and extinction coefficients are determined to be 1.55 and 3.50 cm mmol^{-1}, respectively [2]. Actually, the intensities of the hydroxides at 3,585 and 3,616 cm^{-1} are multiplied by 1.8 and 0.85, respectively, and the sum of them is fitted well to the MS-TPD. Because the ratio of reciprocal of these parameters is close to the ratio of extinction coefficients, the selection of these parameters is supported experimentally. The measured MS-TPD is thus divided into two portions of OH bands from which the number and strength of two Brønsted hydroxides are measured individually.

The procedure for the data analysis is summarized as follows.

1. Band position is corrected, and difference spectra are calculated.
2. Absorption band positions are identified from the difference spectra.
3. IR-TPD for each adsorbed species and OH is calculated.
4. Parameters to cancel the difference in extinction coefficients are determined so as to fit the IR-TPD to the MS-TPD.
5. The number and strength of the acid sites are measured. When the distribution is observed, numbers and strengths of acid sites are measured individually.

Thus not only the number and the strength are measured, but also the structure of acid site is determined. Discrimination between Brønsted and Lewis acidities is possible, and the IR band position of the Brønsted acid sites is measured; such an advantageous characterization technique has not been known yet. Very informative and suggestive characterized data are provided.

3.2.2 H-Y and H-Chabazite

H-Y zeolite has the characteristic that all T sites are equivalent. Thus, four kinds of the Brønsted OH are possible to be stabilized. However, only three kinds of the Brønsted OH were detected by an experiment of the neutron diffraction. Such a simple structure of the zeolite is observed typically on H-Y and H-chabazite. In addition, the H-Y is an industrially important zeolite, because various catalytic reactions are available using the H-Y and H-Y based catalysts. Therefore, a precise characterization of the H-Y must be conducted [3].

Difference spectra obtained after adsorption of ammonia on the H-Y zeolite exhibit profiles almost similar to those on the H-mordenite mentioned above. An additional absorption of ammonia at 1,665 cm^{-1} was observed together with the bending vibration of NH_4^+ at ca. 1,430 cm^{-1}; and this could be identified as the bending vibration of NH_3. Therefore, adsorbed NH_4^+ and NH_3 are identified on the H-Y zeolite. Absorption intensities of the bending vibration of NH_4^+ can be calculated by dividing into three portions at 1,496, 1,430, and 1,369 cm^{-1}, although these are unidentified yet.

Enlarged portion of OH bands shows an interesting profile, as shown in Fig. 3.6. Usually, two kinds of OH are identified in the usual IR spectrum of H-Y zeolite, whereas four kinds of OH band are observed in the IRMS-TPD experiment. Two OH bands observed at high and low wavenumbers are further split into each two OH bands.

3.2 Proton Form Zeolite

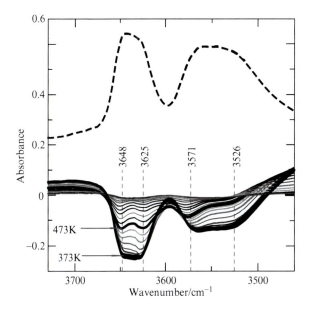

Fig. 3.6 Difference IR spectra of the OH bands obtained on the in situ prepared H-Y zeolite

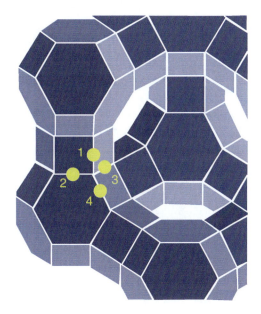

Fig. 3.7 Structure of Y zeolite and four kinds of structurally different oxygen, O1–O4

As shown in Fig. 3.7, the H-Y zeolite has four kinds of Brønsted OH which are structurally different, named O1H–O4H. The protons of O1H and O4H are located in the super cage, that of O2H is in the sodalite cage, and that of O3H is in the double 6-rings. However, the previous neutron diffraction study did not detect the

O4H [4]. Provided that the neutron diffraction study is accepted, only three kinds of OH are detected. Therefore, the band observed at 3,526 cm^{-1} is identified as the O3H due to the previously reported identification [5]. One more OH band at the low wavenumber of 3,571 cm^{-1} is then identified as the O2H in the sodalite cage. On the other hand, the band at 3,648 cm^{-1} is identified as the O1H located at the super cage, and the band at 3,625 cm^{-1} is either due to the O1H with different attached Al atoms or the O4H previously unidentified. Datka et al. report that the distribution of Brønsted OH depends on the number of Al in the nearest next neighbor [6]. One of the OH bands located on the super cage is thus tentatively named as the O1'H.

Intensities of the absorptions are calculated precisely, and the IR-TPD is calculated to be compared with the MS-TPD, as shown in Fig. 3.8. However, the MS-TPD cannot be simply compared with the IR-TPD of adsorbed ammonia and OH. First, by assuming the parameters to cancel the difference in extinction coefficients, sum of IR-TPD of NH_4^+ and NH_3 is fitted to the MS-TPD. The MS-TPD has two desorptions at 430 and 500 K, which correspond to the IR-TPD of NH_3 and of NH_4^+, respectively. Secondly, the IR-TPD summation of three OH bands at 3,648, 3,625, and 3,571 cm^{-1}, and one OH band at 3,526 cm^{-1} are fitted to the IR-TPD of NH_4^+ and of NH_3, respectively.

The l-peak with a high intensity observed in usual ammonia TPD has been assigned to NH_3 adsorbed on NH_4^+, as mentioned in Chap. 2. On the other hand, the so-called l-peak seems to be neglected in the experiment of IRMS-TPD, because the adapted experimental conditions of W/F are so small.

Thus, four OH bands are assigned, and the number and strength of these OH are determined, as shown in Table 3.1. Simultaneously, the extinction coefficients of these OH bands are measured. Careful discussion is however required to fully

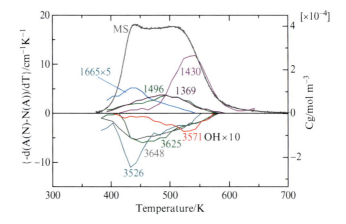

Fig. 3.8 IR-TPD of ammonium cation at 1,496, 1,430, and 1,369 cm^{-1}, ammonia species at 1,665 cm^{-1}, and OH bands at 3,526, 3,648, 3,625, and 3,571 cm^{-1}. For OH, ten times of the value is shown. MS-TPD (C_g) is added for a comparison

3.2 Proton Form Zeolite

Table 3.1 Assignment, number, strength, and extinction coefficient of OH, and accommodating ammonia species on the H-Y zeolite

Band position (cm^{-1})	Assignment	Position	A_0 (mol kg^{-1})	ΔH (kJ mol^{-1})	ε (cm μmol^{-1})	Ammonia species
3,648	O1H	Super cage	0.58	108	1.4	
3,625	O1'H		0.57	110	1.3	NH$_4^+$
3,571	O2H	Sodalite cage	1.1	119	0.52	
3,526	O3H	Double 6-rings	0.8	105	1.3	NH$_3$

understand two different behaviors of OH bands; i.e., these hydroxides are classified into those of OH on which the NH$_4^+$ is adsorbed, and one another with which adsorbed NH$_3$ species interacts. The former three are usual Brønsted OH, and from higher to lower wavenumber the strength of acid site increases and the extinction coefficient decreases. On the other hand, the OH observed at the lowest band position shows the extraordinary character; it shows the smallest value of ΔH for ammonia desorption in spite of the lowest wavenumber, and its intensity changes upon desorption of NH$_3$. Such an unusual behavior may be due to the location of the OH, which is inside of the double 6-ring. Most probably, it is not stabilized as NH$_4^+$ species due to the constraint in cage structure. Thus formed weak interaction of OH with adsorbed NH$_3$ influences on the vibration of OH, and thereby ammonia is desorbed at the low temperature. The weak acidity found by the ammonia TPD is therefore not caused by the intrinsic property of the OH but by the physical structure of the adsorption site.

IRMS-TPD experimental profiles observed on the H-chabazite are explained in this section, because these zeolites show similar profiles based on the similar structure [7]. Difference spectra in the OH region observed on the H-chabazite and thus measured IR-TPD and MS-TPD are shown in Figs. 3.9 and 3.10, respectively. The energy parameter measured by the IRMS-TPD experiment on the chabazite is summarized in Table 3.2.

The H-chabazite also has those four kinds of the Brønsted OH bands, labeled O1H–O4H. Among them, three OHs are located in the 8-ring, and the one in the small pore of 6-ring. The OH band observed at the lowest wavenumber, 3,538 cm^{-1}, is identified as the OH in the small pore, and does not form the NH$_4^+$ cation by the interaction with NH$_3$, as shown in Fig. 3.10. Ammonia molecule is thus readily desorbed at low temperature. The IR band position for the O4H located in the small pore and the derived weak acid strength are similar to those on the O3H in the H-Y zeolite. The OH band named O4H of chabazite should have the Brønsted acidity; however, it does not show the property to accommodate NH$_4^+$ cation, but to interact with NH$_3$; such an unusual behavior is most probably due to the steric constraint in the small pore. Other three OH bands observed from the high- to low wavenumber

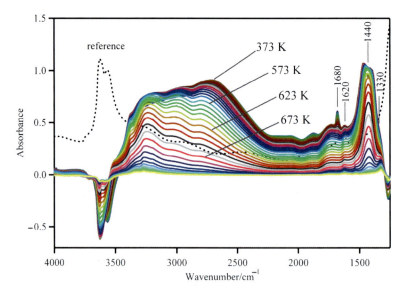

Fig. 3.9 Difference IR spectra in the OH region observed on the H-chabazite

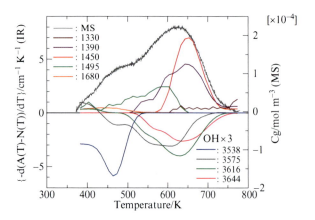

Fig. 3.10 IRMS-TPD and MS-TPD (C_g) on H-chabazite

Table 3.2 Assignment, number, strength, and extinction coefficient of OH, and accommodating ammonia species on the H-chabazite

Band position (cm^{-1})	Assignment	Position	A_0 (mol kg^{-1})	ΔH (kJ mol^{-1})	ε (cm μmol^{-1})	Ammonia species
3,644	O2H		0.50	139	2.9	
3,616	O1H	8-ring	0.65	136	3.2	NH_4^+
3,575	O3H		0.52	133	3.6	
3,538	O4H	6-ring	0.40	105	4.7	NH_3

are identified as O2H, O1H, and O3H, but unlike those on the H-Y zeolite, the acid strength becomes weak in this sequence, as shown in Table 3.2.

Four kinds of OH bands are distinctively observed in the IRMS-TPD experiment, and the distributions of the acid strength are measured on the H-Y and H-chabazite. The dependence of the strength upon the structure is learned from the observation. However, the steric hindrance due to the small pore to suppress the formation of NH_4^+ is required explanation of the observation.

3.3 Modified Zeolites

3.3.1 Multivalent Cation-Modified Zeolite

Multivalent metal cation-exchanged Y zeolites are known to be active for the cracking of hydrocarbons. Divalent alkaline earth cations such as Ca and Ba and trivalent rare earth cation such as La are introduced into the Y zeolite by means of an ion-exchange, and the modified Y zeolites are utilized for the catalytic reaction. Therefore, the characterization of these metal ion-exchanged zeolites is extremely important to understand an effect by these exchanged cations on the enhanced solid acidity. Basic understanding of the cation-exchanged zeolites is also required for the design of active acid catalysts. A method of the IRMS-TPD is applied to these zeolites in order to reveal the enhanced Brønsted acidity [8].

Introduction of Ba and Ca cations to the Y zeolite diminishes the OH band located in the sodalite cage and double 6-rings preferentially. This means that such metal cations as Ba and Ca are exchanged at the cation site which is near the sodalite cage. Remaining OH bands located at the super cage form NH_4^+ upon adsorption of ammonia due to the Brønsted acidity, and the diminished intensity of OH is recovered accompanied by the desorption of ammonia, as shown in Fig. 3.11. The behavior of recovery of the OH band intensity during the TPD experiment shows the presence of two kinds of OH, both located on the super cage. Another fine absorption of adsorbed ammonia is found at 1,661 cm^{-1}, thus showing the presence of the Lewis acid sites. Thus measured changes of the intensity due not only to the adsorbed NH_4^+ and NH_3 but also the OH bands are summarized into the IRMS-TPD profile which could be compared with the MS-TPD, as shown in Fig. 3.12. It is clearly identified that the Brønsted acidity is prevailing on the modified Y zeolites, and the distribution of the sites is outstanding. A strongly enhanced Brønsted acid site is found based on this experiment. Number and strength of two Brønsted acid sites are calculated, and the ΔH for ammonia desorption on the stronger Brønsted OH is plotted against the degree of cation exchange in Fig. 3.13. We understand from this relation that the strength of the Brønsted OH in the super cage is enhanced by the introduction of metal cations: Ca, Ba, and La. On the other hand, a small amount of Lewis acid site is found based on the desorption of ammonia at 430 K, which corresponds to the absorption of NH_3 at 1,665 cm^{-1}.

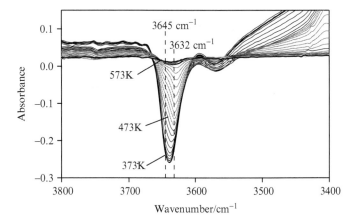

Fig. 3.11 Difference IR spectra in the OH region, which is recovered due to the desorption of ammonia in the BaHY

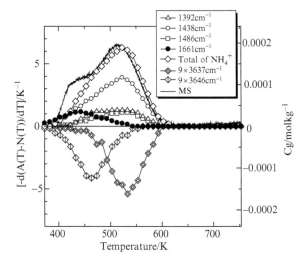

Fig. 3.12 IRMS-TPD profile compared with the MS-TPD on BaHY

The role of the cations for the enhancement of the acidity of Brønsted OH is an interesting subject. Based on the findings of IRMS-TPD, it is found that the multivalent metal cations located on the small pore play the role of enhancing the acidity of Brønsted OH located in the large pore. The electron accepting, i.e., the Lewis acidic property of the cation is not outstanding. Thus, the mechanism most probably estimated is as follows, i.e., first, the cation withdraws electron from the OH to increase the partial charge of H^+; and thus the strength of Brønsted OH is enhanced. The property of the cation as an electron acceptor leads to the strong Brønsted acidity. Quantitative study is possible by applying the DFT calculation to this system, which is mentioned in Chap. 4.

3.3 Modified Zeolites

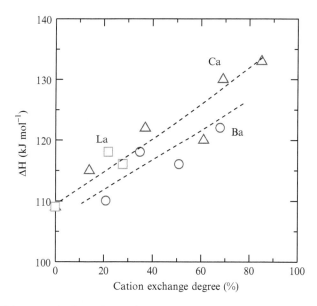

Fig. 3.13 ΔH for ammonia desorption on the stronger Brønsted OH plotted against the degree of cations Ba, Ca, and La exchange

3.3.2 Ultrastable Y (USY) Zeolite

Ultrastable Y (USY) zeolite has been utilized for cracking of hydrocarbons in the petroleum refinery process. Therefore, the solid acidity which would play an important role in the catalytic reaction has to be fully studied. Many studies from various view-points have already been performed. It has been found that the steaming of Y zeolite creates the mesoporosity, and produces the extra-framework Al. Although such modifications of structure and composition are found clearly, these effects on the solid acidity remain unsolved. Because of the complex profiles of the acidity and structure, it is not easy to reveal the solid acidity sufficiently. Therefore, it is interesting to apply the method of IRMS-TPD to the study on USY zeolite.

In the present chapter, the EDTA (ethylenediaminetetraacetic acid)-treated USY zeolite is used as an example of the material. USY is first prepared from ammonium Y zeolite by steaming at a high temperature, for example 823 K. Thus obtained USY is further treated by Na_2H_2–EDTA. Al atoms in the framework are released from the zeolite lattice in the first step (steaming), and the locations and environments of Al atoms outside of the framework are changed in the next step (treatment in a chelating agent). Control of the positions of extra-framework Al atoms is a key issue to create the strong solid acidity. Treatments of USY with acidic and basic solutions provide various phenomena such as further removal of the framework Al, washing out of the extra-framework Al, and/or reinsertion of the extra-framework Al. Such multiple changes cause a controversy over the origin of high activity of USY. Unlike the acidic and basic solutions, a chelating agent with weak basicity

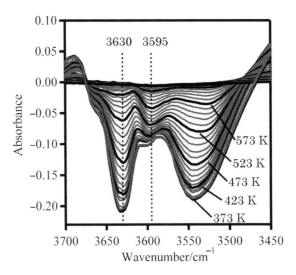

Fig. 3.14 Difference IR spectra in the OH region observed on the USY zeolite

(Na$_2$H$_2$–EDTA) mainly changes the location of extra-framework Al, resulting in the following properties. Experimental conditions for the preparations are optimized [9], and the most adequately prepared USY is used in the measurements of the solid acidity.

Difference spectra in the OH region of USY are shown in Fig. 3.14 [10]. Negative observation is due to the decrease of the intensity, and the Brønsted acidities of these OH bands are observed. Two large negative changes of the OH at high- and low wavenumbers are due to the differences in intensities of two kinds of OH bands in the Y zeolite. Unlike on the H-Y zeolite shown above, these OH band intensities do not split further. However, in addition, one more OH band is observed at 3,595 cm^{-1} between two inherent OH bands. The OH band newly detected in the present study has already been discerned to be ascribed to the active acid sites by Lunsford et al. [11]. Assumed that there are three kinds of OH bands, IRMS-TPD profiles are calculated as shown in Fig. 3.15. As in the preceding explanation, the MS-TPD is well fitted to the IR-TPD of NH$_4$$^+$ to prove the predominant distribution of Brønsted acidity. The Brønsted acidity consists of three OHs at 3,630, 3,545, and 3,595 cm^{-1}, which show different thermal behaviors. The procedure mentioned above is applied to the calculation of the number and the strength of three Brønsted OH bands, as shown in Table 3.3. The OH clearly detected at 3,595 cm^{-1} has the ΔH value of 137 kJ mol^{-1}. The value of ΔH corresponds to the strong acidity comparable to the HZSM-5. Most probably, this is the strong Brønsted acid site active for the cracking of hydrocarbons. It is important to notice that the OH band, which is covered by the large and broad absorptions of other bands, is discovered by using a technique of TPD experiment.

3.4 Distribution of Brønsted Acid Sites Dependent on the Concentration

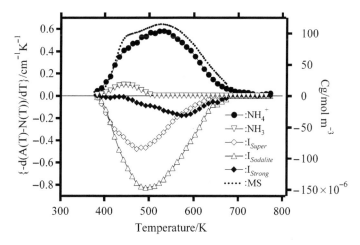

Fig. 3.15 IRMS-TPD profiles and MS-TPD (C_g) on the USY zeolite

Table 3.3 Assignment, number, strength, and extinction coefficient of OH, and accommodating ammonia species on the H-Chabazite

Band position (cm^{-1})	Assignment	A_0 (mol kg^{-1})	ΔH (kJ mol^{-1})	ε (cm μmol^{-1})
3,635	In super cage	0.39	116	2.9
3,595	Created	0.47	137	3.2
3,540	In sodalite cage	0.27	122	4.7

The present method of IRMS-TPD unveils the important character of the acidity in USY. Combined with other studies, e.g., NMR observation and DFT (density functional theory) calculation, further clear characterization is undertaken to reveal the structure and its relation with the strong Brønsted acidity, as shown in Chap. 4.

3.4 Distribution of Brønsted Acid Sites Dependent on the Concentration

Brønsted acidities on the proton form mordenite, chabazite, and Y zeolite, i.e., zeolite species with ca. 100% exchange degree with proton, have been elucidated in previous sections. It is interesting that two to four kinds of acid sites are clearly revealed by utilizing an advantageous property of the IRMS-TPD experiment. In the next step, the distribution of these acid sites is studied with varying the degree of exchange on the NaH-type mordenite and chabazite. The stability of the acid site which depends on the structure or the site location is studied.

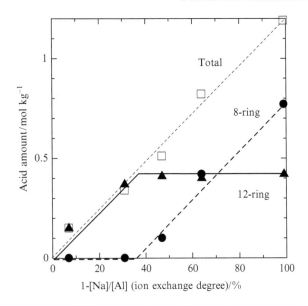

Fig. 3.16 Plots of numbers of OH groups in 12- (*solid line*) and 8-rings (*dotted line*) against the ion exchange degree of ammonium cation

As shown in the previous sections, the IR-TPD spectra of Brønsted OH groups at 3,585 and 3,616 cm^{-1} on H-mordenite are multiplied by 1.8 and 0.85, respectively, and the sum of them is fitted well to the MS-TPD; thus, the numbers of two kinds of OH groups are quantitatively measured. By means of this method, the OH groups on NaH–mordenite samples with different ion exchange degrees (different Na/Al ratios) are quantified. Figure 3.16 shows the numbers of acidic OH groups in 12- and 8-rings against the ion exchange degree of proton/ammonium cation [1]. At a low exchange degree of proton, only the acid site in 12-ring is detected. In the region between 0 and 40% of the ion exchange degree, the number of acidic OH groups in 12-ring linearly increases with increasing the number of H(= Al–Na). In this region, the Na cation is preferentially located in 8-ring; and in other words, the proton is located preferentially in 12-ring. The number of OH in 12-ring increases up to ca. 0.4 mol kg^{-1}, and further increase in the ion exchange generates the acidic OH located in 8-ring. The OH in 8-ring increases linearly with increasing the exchange degree, and the concentration finally arrives at ca. 0.8 mol kg^{-1} on the fully exchanged H-mordenite. This indicates that, at least on this sample (in situ H-mordenite prepared from a reference catalyst JRC-Z-M15 (Na-form)), about two-thirds of the Al atoms are located at T_3 position.

The Brønsted acid strengths of the two kinds of OH groups are also quantified. The heats of ammonia adsorption on the OH groups in 12- and 8-rings are ca. 140 and 150 kJ mol^{-1}, respectively, indicating the stronger acidity of the Brønsted OH located in the 8-ring [1].

Also in the case of Y zeolite, the change in the numbers of OH groups with varying the ion exchange degree is studied, as shown in Fig. 3.17. The Brønsted OH

3.5 Relationship Between Stretching Frequency and Ammonia Desorption

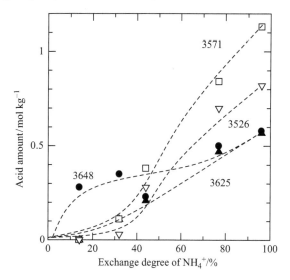

Fig. 3.17 Amounts of four kinds of OH in H-Y zeolite observed at 3,648 (O1H), 3,625 (O1′H), 3,571 (O2H), and 3,526 cm^{-1} (O3H) plotted against the ion exchange degree

stabilized at a small exchange degree is the O1H observed at 3,648 cm^{-1}, and other three kinds of OH appear above ca. 30% of the exchange degree. The O1H at the highest wavenumber is most readily stabilized.

3.5 Relationship Between Stretching Frequency and Ammonia Desorption Heat of OH Group

When the averaged heat of ammonia desorption obtained by the conventional MS-TPD method is plotted against the IR band position measured by individual IR experiments, no relation is observed between them, as shown in Fig. 3.18. The improved IRMS-TPD method, however, provides us with new information. The quantification of distribution of acid sites in MOR, FER, and FAU gives a new insight into the interpretation of the acid strength, as mentioned below.

- On MOR, the high averaged desorption heat is largely owing to the stronger acid site in 8-ring which shows a shoulder of IR band, while the major part of IR band is due to the weaker acid site in 12-ring.
- On FER, the acid sites in 10- and 6-rings are distinguished.
- On the cation-exchanged Y zeolite and USY, the OH band whose acid strength is enhanced by the multivalent cation is identified.

With this information, the revised relation is shown in Fig. 3.19. In this figure, the acid sites located in oxygen member rings larger than 8-rings are shown by the filled symbols, and those in 6-rings are shown by the open symbols. A correlation is found between the ammonia desorption heat and the wavenumber (therefore, frequency of the OH stretching vibration) in which the lower wavenumber gives the

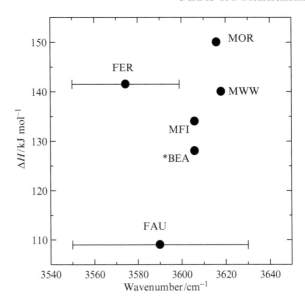

Fig. 3.18 Plots of ΔH against IR band position based on individual measurements of MS-TPD and IR

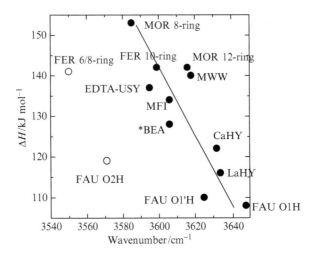

Fig. 3.19 Relation between ΔH and IR band position based on IRMS-TPD measurements

stronger acidity. The low frequency must show the weak OH bond and/or the large OH distance, corresponding to the high proton-donating ability to a basic molecule, namely, the strong Brønsted acidity.

This correlation is obtained on acid sites in the pores larger than 8-ring, while the acid sites in 6-ring show the weaker acidities than those predicted from their low wavenumbers. The OH frequency in the 6-ring is lowered by hydrogen bond(s) between H in the OH group and O in the pore wall due to the small pore size, and/or the adsorption of ammonia on the OH is structurally hindered by the small pore, resulting in the low desorption heat.

3.6 Distorted Structure of Zeolite and Related Material with Lewis Acidity and Broad Distribution of Acid Strength

In this section, some examples of Lewis acid sites observed on zeolites are explained. As described in the previous sections, the framework Al generates Brønsted acid site. Lewis acidity is observed on zeolites with defects, amorphous parts, and/or extra-framework Al species.

Recently, novel synthesis of mesoporous zeolite, a zeolite with mesoporosity or large external surface area, is being attempted by many researchers. Newly developed zeolites often have distorted structures on the initial stage of development. Therefore, analysis of physical structure is carried out at first. Then, because the most important purpose of these studies is to obtain a highly active solid acid catalyst, the analysis of acidic property should be important.

Ryoo et al. develop a synthesis method of mesoporous ZSM-5 zeolite using an organosilane SDA (structure directing agent) [12]. They first developed a method in which tetraethoxysilane as a silicon source and the SDAs were simultaneously mixed (hereafter, method 1). The IR spectra of a sample prepared by method 1 are shown in Figs. 3.20 and 3.21. In the low wavenumber region (Fig. 3.20), in addition to the NH_4^+ band ($1,448\,cm^{-1}$), a band is observed at $1,321\,cm^{-1}$. A symmetric deformation vibration of NH_3 bounded to a metal cation must show a band around $1,200\,cm^{-1}$ [13], and the skeletal vibration of silicate is considered to diminish major part of the deformation band. The $1,321\,cm^{-1}$ band is believed to be the remaining part of $1,200\,cm^{-1}$ band, and thus attributable to NH_3 coordinated to Lewis acid site. The overlapping bands of NH_4^+ and NH_3 are deconvoluted, as shown in Fig. 3.20b.

On the other hand, in the high wavenumber region (Fig. 3.21), OH on dislodged Al ($3,781\,cm^{-1}$), isolated Si–OH ($3,746\,cm^{-1}$) and Al–OH ($3,670\,cm^{-1}$) are detected as well as the acidic SiOHAl ($3,607\,cm^{-1}$). The overlapping spectrum is divided into five portions as shown in Fig. 3.21b, with an assumption of additional fraction of SiOH with an interaction ($3,738\,cm^{-1}$).

IR-TPD spectra of these bands are shown in Fig. 3.22 with overlapping the MS-TPD. In this figure, the IR-TPDs of NH_4^+ (IR-TPD of Brønsted acid site) and of NH_3 (IR-TPD of Lewis acid site) are shown. The IR-TPDs have been magnified by selecting a pair of coefficients[1] to obtain the best fitting between MS-TPD

---

[1] The coefficients should be in inversely proportional to the molar extinction coefficients of NH_4^+ and NH_3, but in this case, they are disturbed by the skeletal vibration of silicate.

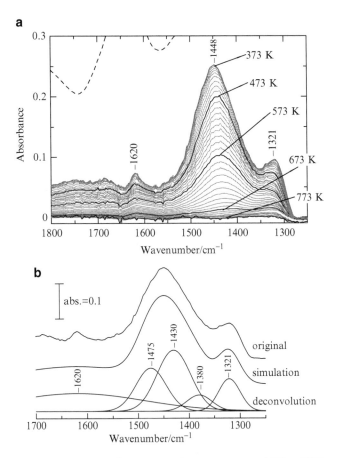

Fig. 3.20 (a) The 1,250–1,800 cm^{-1} region of difference IR spectra $A(T) - N(T)$ caused by adsorption and desorption of ammonia in the TPD experiment on mesoporous H-ZSM-5 synthesized by method 1 (see text). (b) An example of the deconvolution

and the sum of IR-TPDs. The MS-TPD and the sum of IR-TPDs are well fitted at 500–700 K, and thus Brønsted and Lewis acid sites are individually quantified.

The IR-TPD of SiOHAl (3,607 cm^{-1}) is close to the mirror image of the IR-TPD of Brønsted acid site, showing that the Brønsted acid site is mainly ascribed to the framework SiOHAl group. However, thermal behaviors other OH groups[2] are not related with the TPD of Brønsted acid site. The number of the Brønsted acid sites (not shown) is much lower than the Al atoms. Considerable amount of Lewis acid sites is found. In addition, the MS-TPD shows a peak at 450 K, where neither a

---

[2] These OH groups are not Brønsted acid sites, but the stretching vibration of OH is affected by the adsorption of ammonia. Probably the adsorption changes the vibration frequency through such a reaction as (continued on next page)

3.6 Distorted Structure of Zeolite and Related Material

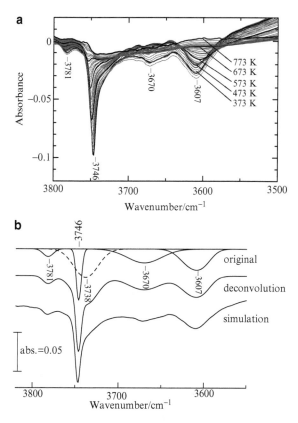

Fig. 3.21 (a) The 3,500–3,800 cm^{-1} region of difference IR spectra $A(T) - N(T)$ caused by adsorption and desorption of ammonia in the TPD experiment on mesoporous H-ZSM-5 synthesized by method 1. (b) An example of the deconvolution

---

(continued from previous page)

where the dotted line between O and H shows the weakened OH bond.
Among these OHs, the isolated SiOH (3,746 cm^{-1}) seems to be related with relatively strong Lewis acid site. Probably a defect with such an SiOH modifies the acidic property of adjacent SiOHAl. On the other hand, the AlOHs (3,781 and 3,670 cm^{-1}) are related with both of the Lewis acid site (1,320 cm^{-1} band) and undetected species showing the MS-TPD peak at 450 K; the latter is possibly related with the 1,600 cm^{-1} band. The precise attribution of these bands has not been clear.

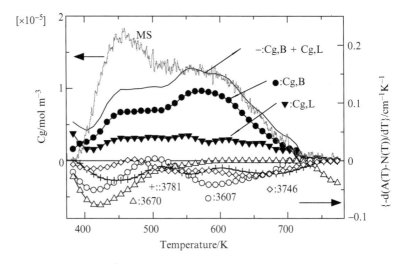

Fig. 3.22 IR-TPD for NH_4^+ (C_g, B), NH_3 (C_g, L), and OH groups (wavenumbers are shown in figure) to be compared with MS-TPD on mesoporous H-ZSM-5 synthesized by method 1

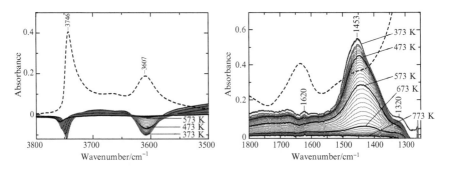

Fig. 3.23 Difference IR spectra $A(T) - N(T)$ caused by adsorption and desorption of ammonia in the TPD experiment on mesoporous H-ZSM-5 synthesized by method 2

distinct desorption of NH_4^+ nor NH_3 is observed in the IR-TPD. It is believed that only a small fraction of Al atoms contributes to generate the Brønsted acid site in this sample.

Based on this information, the synthesis method is improved. In place of tetraethoxysilane, water glass is used as a silicon source, and the mixing procedure is improved; after mixing the silicon and aluminum sources, the SDA is added (method 2), in order to disperse Al in the silicate matrix. Figures 3.23 and 3.24 show the spectra of a sample prepared by method 2. Figure 3.23 shows the decrease of Lewis acid sites ($1,320\,cm^{-1}$), extra-framework AlOH ($3,781$ and $3,670\,cm^{-1}$) in the improved synthesis method compared to method 1. Figure 3.24 shows a large amount of Brønsted acid sites and a small amount of Lewis acid sites, thus justifying the presence of the acid site in the ZSM-5 framework.

3.6 Distorted Structure of Zeolite and Related Material

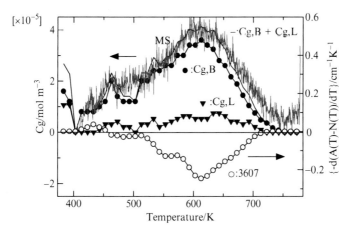

Fig. 3.24 IR-TPD for NH_4^+ (C_g, B), NH_3 (C_g, L) and OH group (wavenumber is shown in figure) to be compared with MS-TPD on mesoporous H-ZSM-5 synthesized by method 2

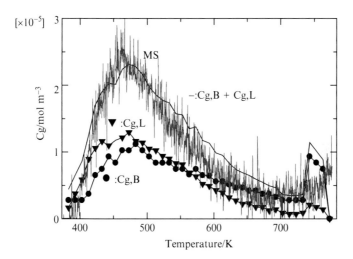

Fig. 3.25 IR-TPD for NH_4^+ (C_g, B), NH_3 (C_g, L) to be compared with MS-TPD on Al-MCM-41

Examples of IRMS-TPD spectra on amorphous aluminosilicates can be seen in Figs. 3.25 and 3.26. The former shows the spectrum on Al-MCM-41, a mesoporous aluminosilicate with an amorphous microstructure. The latter is that of a commercially available silica alumina catalyst also with an amorphous structure. The acidic properties of these amorphous materials are summarized as follows [14].

- Considerable amount of Lewis acid sites is observed.
- Distributions of acid strengths of both of Lewis and Brønsted acid sites are broad.

The former indicates that defect and extra-framework Al generate Lewis acidity. The latter tells us that the acid strength is controlled by bond angles and lengths

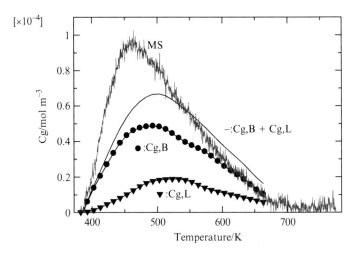

Fig. 3.26 IR-TPD for NH_4^+ (C_g, B), NH_3 (C_g, L) to be compared with MS-TPD on amorphous silica–alumina Nikki N-631L (a commercially available sample)

around the acid site, and hence it is widely distributed in the amorphous materials where the bond angles and lengths are not strictly limited unlike pure crystals.

As shown above, the newly synthesized mesoporous ZSM-5 has considerable amount of Lewis acidity. This is presumably due to the presence of defects and extra-framework Al. On an ex situ H-β zeolite, similar Lewis acidity is observed. Steaming of the zeolite at a high temperature decreases the Brønsted acid sites and increases the Lewis acid sites [15].

3.7 Measurements of Metal Oxide Overlayer

Loading of acidic oxides on basic metal oxides provides us various kinds of solid acid catalysts (Table 3.4). Generally, a monolayer of the acid promoter spreads over the surface, and the catalytic activity is generated where the monolayer covers the surface, suggesting the generation of acid site on the monolayer. WO_3/ZrO_2 and SO_4^{2-}/ZrO_2 are representatives of these monolayer-type solid acids. They are believed to possess very strong acidity, because such a difficult acid-catalyzed reaction as alkane skeletal isomerization proceeds on them [16]. Some of these oxides are known to be oxidation or reduction catalysts; for example, MoO_3/SnO_2 catalyzes a partial oxidation of methanol into formaldehyde, and V_2O_5/TiO_2 catalyzes a selective reduction of NO with NH_3. For these red–ox reactions, the acid site (especially Brønsted acid site) is believed to play a role of active site or promoter.

The distribution of acid strength on such a combined metal oxide, similarly to the amorphous aluminosilicate shown in the previous section, is generally broad compared to that on zeolites. Brønsted and Lewis acid sites always coexist, and

3.7 Measurements of Metal Oxide Overlayer

Table 3.4 Combination of acidic oxide and basic metal oxide support forming a monolayer of acidic oxide on the surface and acid site on it

Promoter	Support	Responsible reaction	Ref.
SO_4^{2-}	ZrO_2	Skeletal isomerization of alkane, Friedel-Crafts type alkylation	[16, 17]
WO_3	ZrO_2	Alkylation of alkane by alkane, alkylation of aromatics	[18]
WO_3	TiO_2	Hydration of ethene into ethanol	[19]
MoO_3	ZrO_2	Selective oxidation of methanol into formaldehyde	[20, 21]
MoO_3	SnO_2	Selective oxidation of methanol into formaldehyde	[22, 23]
V_2O_5	TiO_2	Selective reduction of NO with NH_3	[24]
V_2O_5	Al_2O_3		[25]
V_2O_5	ZrO_2		[26]
V_2O_5	SnO_2		[26]
SiO_2	Al_2O_3	Double bond isomerization of butane	[27]
SiO_2	TiO_2	Double bond isomerization of butane	[28]
SiO_2	ZrO_2	Double bond isomerization of butane	[28]

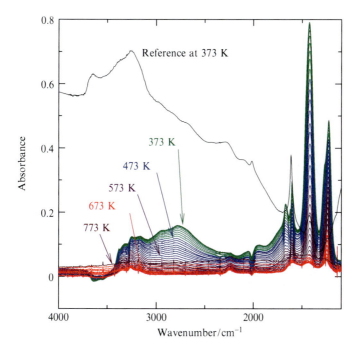

Fig. 3.27 Difference IR spectra of ammonia adsorbed on WO_3/TiO_2 with a reference spectrum

therefore, individual quantification of these acid sites is important. The IRMS-TPD method is a powerful tool for analysis of the acidic property of the combined metal oxide.

Figure 3.27 shows IR spectra of ammonia adsorbed on WO_3/TiO_2 with $6\,nm^{-2}$ of W atom density on the surface. Bending bands of NH_4^+ ($1,430\,cm^{-1}$) and NH_3

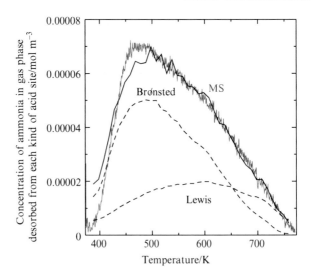

Fig. 3.28 Curve fitting of IR- and MS-TPD on WO_3/TiO_2

($1{,}220\,cm^{-1}$) are observed after the adsorption of ammonia at 373 K, and their intensities gradually decrease with heating the catalyst. From the rates of decrease in intensities of these bands, IR-TPD spectra are calculated for Brønsted and Lewis acid sites. Coefficients are searched in order to fit the sum of IR-TPDs of Brønsted and Lewis acid sites with MS-TPD. The curve fitting is well carried out in this case, as shown in Fig. 3.28. Thus, TPD spectra of Brønsted and Lewis acid sites are calculated. From the peak intensity, position and shape, the number and strength (ammonia adsorption heat) are determined for each type of acid site [19].

The acid strength of the combined metal oxide has a broad distribution unlike those of zeolites, as imagined from the broad TPD peaks. We have developed a method to calculate the distribution of ammonia desorption heat. Fraction ($f_{\Delta H}$) acid sites with different ΔH are assumed. The ΔH is assumed at every $5\,kJ\,mol^{-1}$ from 90 to $250\,kJ\,mol^{-1}$. The acid amount is assumed to be the total acid amount. The desorption profile is simulated based on (3.2).

$$C_g = -\frac{\beta A_0 W}{F}\frac{d\theta}{dT} = \frac{\theta}{1-\theta}\frac{P^0}{RT}\exp\left(-\frac{\Delta H^0}{RT}\right)\exp\left(\frac{\Delta S^0}{R}\right). \quad (3.2)$$

A TPD spectrum is assumed to be the sum of these fractions with suitable set of coefficients (A_{90}–A_{250}) as follows:

$$\text{TPD} = A_{90}f_{90} + A_{95}f_{95} + \cdots + A_{250}f_{250}. \quad (3.3)$$

The set of coefficients (A_{90}–A_{250}) is searched to fit thus simulated TPD with the experimentally observed IR-TPD of Brønsted or Lewis acid site. The numerical searching is carried out using a solver function of Microsoft Excel. The set is

3.8 Extinction Coefficients of NH_4^+ and NH_3

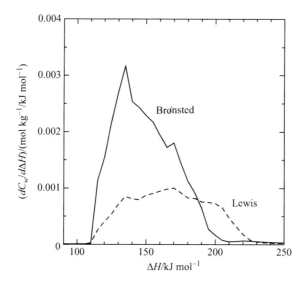

Fig. 3.29 Distribution of acid strength (ammonia adsorption heat) of Brønsted and Lewis acid sites on WO_3/TiO_2

selected based on a minimum square method. Figure 3.29 shows the distribution of ΔH thus determined from TPD spectra shown in Fig. 3.28.

Hydration of ethene (ethylene) into ethanol is performed using an H_3PO_4/SiO_2 catalyst, and this catalyst causes environmental problem due to eluted phosphoric compounds. Replacement of this catalyst by insoluble solid acid catalyst has been demanded. The acidic property of WO_3/TiO_2 thus measured well explains the catalytic activity for this reaction; the activity is dependent on strong Brønsted acidity. As a result of application of the ammonia IRMS-TPD method to this system, a promising candidate, WO_3 monolayer loaded on TiO_2 has been found [19].

The acidic property of SO_4^{2-}/ZrO_2 has also been analyzed. The effect of calcination temperature of support zirconia is clarified; the strong Brønsted acidity is generated when the support is uncalcined or calcined at a low temperature before loading of sulfate species [29]. An example of TPD spectra of Brønsted and Lewis acid sites on SO_4^{2-}/ZrO_2 is shown in Fig. 3.30.

3.8 Extinction Coefficients of NH_4^+ and NH_3 Adsorbed on Brønsted and Lewis Acid Sites, Respectively

To determine the extinction coefficient, the exact equation correlating the MS-TPD and IR-TPD is derived. From the material balance in the system,

$$-\beta C_w W \frac{d\theta}{dT} = FC_g, \qquad (3.4)$$

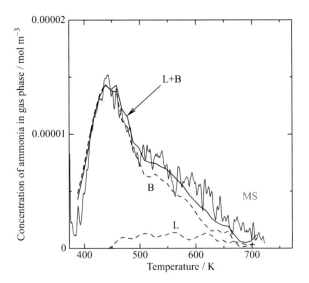

Fig. 3.30 TPD spectra of Brønsted (B) and Lewis (L) acid sites on SO_4^{2-}/ZrO_2

where β, C_w, W, θ, T, F, and C_g are ramp rate (K s^{-1}), number of the discussed type of adsorption site (mol kg^{-1}), sample weight (kg), coverage, temperature (K), flow rate of carrier (m^3 s^{-1}), and concentration of desorbed gas in the gas phase (mol m^{-3}), respectively, as shown in Chap. 2.

On the other hand, Lambert–Beer law states

$$A = \varepsilon c l, \tag{3.5}$$

where A, ε, c, and l are absorbance, molar extinction coefficient, concentration (mol m^{-3}), and beam length (m), respectively. If the absorbance A is expressed by the peak height in an IR spectrum, A is a dimensionless parameter, and the unit of ε is m^2 mol^{-1}. Or, if A is expressed by the peak area, the units of A and ε are m^{-1} (as well as the wavenumber) and m mol^{-1}, respectively.

Here, a sample wafer is assumed to have a cross-sectional area S (m^2) and a thickness l (m). The volume of wafer v (m^3) is

$$v = Sl. \tag{3.6}$$

When all the adsorption sites are occupied, the concentration of an adsorbed compound c in this wafer is

$$c = \frac{C_w W}{v} = \frac{C_w W}{Sl}. \tag{3.7}$$

From (3.5) and (3.7),

$$A = \frac{\varepsilon C_w W}{S}. \tag{3.8}$$

3.8 Extinction Coefficients of NH_4^+ and NH_3

The IR-TPD, namely, the rate of decrease in IR band intensity, $-dA/dT$, is correlated with the decrease of coverage $-d\theta/dT$ as

$$-\frac{dA}{dT} = -\frac{\varepsilon C_w W}{S}\frac{d\theta}{dT}. \quad (3.9)$$

Comparison of (3.4) and (3.9) gives

$$C_g = \frac{\beta S}{\varepsilon F}\left(-\frac{dA}{dT}\right). \quad (3.10)$$

This tells us that the MS-TPD (C_g) is a product of a constant determined by the experimental conditions ($\beta S/F$), a reciprocal of ε, and the IR-TPD ($-dA/dT$).

From the IRMS-TPD experiments, one can determine the molar extinction coefficients of $1{,}450\,\text{cm}^{-1}$ band (NH_4^+ bound to Brønsted acid site), $1{,}200\,\text{cm}^{-1}$ band (NH_3 coordinated to Lewis acid site), and various OH groups by two procedures.

- Based on (3.10), the coefficient $\beta S/\varepsilon F$ can be obtained from the fitting of MS- and IR-TPDs as shown in Fig. 3.28. From this value, ε is determined.
- Based on (3.8), the IR peak intensity at 373 K is proportional to $C_w W/S$, and the slope should be ε, if ε is a constant on various samples with different C_w.

Theoretically both methods give the same answer, but practically experimental errors affect the results. Here the latter method is shown.

In the experiments dealt in this chapter, the diameter of wafer is 1 cm and therefore S is fixed to be $7.9 \times 10^{-5}\,\text{m}^2$. Figure 3.31 shows the plots of A of $1{,}450\,\text{cm}^{-1}$

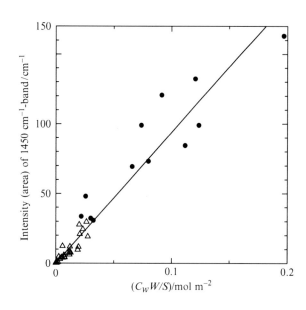

Fig. 3.31 Relationship between absorbance of $1{,}450\,\text{cm}^{-1}$ band (NH_4^+ on Brønsted acid site) at 373 K and $C_w W/S$, where C_w is the number of Brønsted acid site, on zeolites (*filled circle*) and non-zeolitic solid acids (*open triangle*: TiO_2, ZrO_2, Al_2O_3, WO_3/TiO_2, WO_3/ZrO_2, WO_3/Al_2O_3, MoO_3/TiO_2, MoO_3/ZrO_2, MoO_3/Al_2O_3, and V_2O_5/TiO_2)

band (NH_4^+ on Brønsted acid site) at 373 K against $C_w W/S$ on various samples, i.e., zeolites and non-zeolitic materials. Although there are large deviations probably due to experimental errors, a proportional relationship is found between A and $C_w W/S$. The slope, ε, is thus determined to be 938 $cm^{-1}\ m^2\ mol^{-1}$.

The above analysis is based on the peak intensities expressed by peak areas, and errors become large when the peak height is adopted [30]. Thus determined extinction coefficient of deformation vibration of NH_4^+ can be compared to literature, but only the values based on the peak height has been known. Datka et al. reported a value of coefficient based on the peak height; 14.7 $m^2\ mol^{-1}$ on mordenite [31]. This is reasonably in agreement with our study on mordenite (13.7 $m^2\ mol^{-1}$) [1]. These findings support the validity of IRMS-TPD measurements, but simultaneously, the largely scattered relations in Fig. 3.31 indicate that the quantification of acid sites only from the IR peak intensity is not easy. However, it is possible to measure approximately the amount of Brønsted acid site only from the IR absorption intensity based on this averaged value.

Figure 3.32 shows plots of A of 1,200 cm^{-1} band (NH_3 on Lewis acid site) at 373 K against $C_w W/S$. The coefficient is determined to be 725 $cm^{-1}\ m^2\ mol^{-1}$. This parameter of Lewis acid site has been determined for the first time in this study.

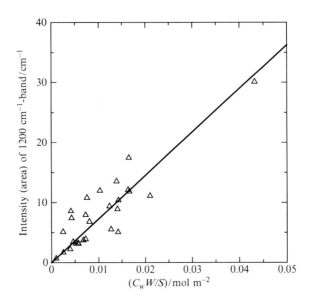

Fig. 3.32 Relationship between absorbance of 1,200 cm^{-1} band (NH_3 on Lewis acid site) at 373 K and $C_w W/S$, where C_w is the number of Lewis acid site, on TiO_2, ZrO_2, Al_2O_3, WO_3/TiO_2, WO_3/ZrO_2, WO_3/Al_2O_3, MoO_3/TiO_2, MoO_3/ZrO_2, MoO_3/Al_2O_3, and V_2O_5/TiO_2

References

1. M. Niwa, K. Suzuki, N. Katada, T. Kanougi, T. Atoguchi, J. Phys. Chem. B **109**, 18749 (2005)
2. J. Datka, B. Gil, A. Kubacka, Zeolites **17**, 428 (1996)
3. K. Suzuki, N. Katada, M. Niwa, J. Phys. Chem. C **111**, 894 (2007)
4. M. Czjzek, H. Jobic, A.N. Fitch, T. Vogt, J. Phys. Chem. **96**, 1535 (1992)
5. F.R. Sarria, O. Marie, J. Saussey, M.J. Daturi, J. Phys. Chem. B **109**, 1660 (2005)
6. J. Datka, B. Gil, J. Catal. **145**, 372 (1994)
7. K. Suzuki, G. Sastre, N. Katada, M. Niwa, Phys. Chem. Chem. Phys. **9**, 5980 (2007)
8. T. Noda, K. Suzuki, N. Katada, M. Niwa, J. Catal. **259**, 203 (2008)
9. N. Katada, Y. Kageyama, K. Takahara, T. Kanai, H.A. Begum, M. Niwa, J. Mol. Catal. A **211**, 119 (2004)
10. M. Niwa, K. Suzuki, K. Isamoto, N. Katada, J. Phys. Chem. B **110**, 264 (2006)
11. P.O. Fritz, J.H. Lunsford, J. Catal. **118**, 85 (1989)
12. M. Choi, H.S. Cho, R. Srivastava, C. Venkatesan, D.-H. Choi, R. Ryoo, Nat. Mater. **5**, 718 (2006)
13. K. Nakamoto, *Infrared and Raman Spectra of Inorganic and Coordination Compounds*, 3rd edn. (Wiley, New York, 1978), pp. 199
14. K. Suzuki, Y. Aoyagi, N. Katada, M. Choi, R. Ryoo, M. Niwa, Catal. Today **132**, 38 (2008)
15. M. Niwa, S. Nishikawa, N. Katada, Microporous Mesoporous Mater. **82**, 105 (2005)
16. K. Arata, Adv. Catal. **37**, 165 (1990)
17. N. Katada, J. Endo, K. Notsu, N. Yasunobu, N. Naito, M. Niwa, J. Phys. Chem. B **104**, 10321 (2000)
18. N. Naito, N. Katada, M. Niwa, J. Phys. Chem. B **103**, 7206 (1999)
19. N. Katada, Y. Iseki, A. Shichi, N. Fujita, I. Ishino, K. Osaki, T. Torikai, M. Niwa, Appl. Catal. A Gen. **349**, 55 (2008)
20. Y. Matsuoka, M. Niwa, Y. Murakami, J. Phys. Chem. **94**, 1477 (1990)
21. N. Narishige, M. Niwa, Catal. Lett. **71**, 63 (2001)
22. M. Niwa, H. Yamada, Y. Murakami, J. Catal. **134**, 331 (1994)
23. M. Niwa, J. Igarashi, Catal. Today **52**, 71 (1999)
24. M. Niwa, Y. Matsuoka, Y. Murakami, J. Phys. Chem. **91**, 4519 (1987)
25. M. Niwa, S. Inagaki, Y. Murakami, J. Phys. Chem. **89**, 3869 (1985)
26. Y. Habuta, N. Narishige, K. Okumura, N. Katada, M. Niwa, Catal. Today **78**, 131 (2003)
27. N. Katada, T. Toyama, M. Niwa, J. Phys. Chem. **98**, 7647 (1994)
28. M. Niwa, N. Katada, Y. Murakami, J. Catal. **134**, 340 (1992)
29. N. Katada, T. Tsubaki, M. Niwa, Appl. Catal. A Gen. **340**, 76 (2008)
30. K. Suzuki, T. Noda, N. Katada, M. Niwa, J. Catal. **250**, 151 (2007)
31. J. Datka, B. Gil, A. Kubacka, Zeolites **15**, 501 (1995)

Chapter 4
DFT Calculation of the Solid Acidity

Abstract Density functional theory (DFT) is applied to calculate the ammonia desorption energies on zeolites with various framework types and Y zeolites with various exchange cations. The calculated desorption energy is in good agreement with the ammonia IRMS-TPD measurements. Relationships between the acid strength and geometric parameters are found.

4.1 DFT Calculation

4.1.1 DFT Calculation Applied to the Study on Brønsted Acidity

Density functional theory (DFT) calculation is a method for the quantum chemical study, which has recently been utilized most frequently. Its applications are very broad and directed to every field of science and engineering. By applying the theoretical calculations, deep insights into the research subject are given. It may be possible to change the viewpoints of the investigation dramatically. Dr. Kohn who is a Nobel laureate for developing the DFT method, in his Nobel lecture in 1999, first showed the structure of methanol included in the cavity of sodalite as an example of the calculation [1]. Zeolite is a fine crystal consisting of a simple structure, and the adsorption of a simple molecule methanol may be a good example to show the significance of the DFT calculation. We learn from the historical event that the adsorption of ammonia on the H-type zeolite also is a subject to which the DFT is applied most adequately. The study on the solid acidity of zeolite by a method of DFT not only supports the experimental observation by the ammonia TPD, but also leads us to understanding the acidity of zeolite on an atomic level. IRMS-TPD experiment, described above, reveals clearly the acidity on Brønsted sites, individually. Accumulation of such precise acidity profiles of zeolites provides us a good chance for the theoretical investigation, because a comparison of the theoretical calculation with the experimental observation is possible. A combination of the DFT calculation with the IRMS-TPD experiment becomes a powerful method to characterize the acidity.

In the DFT study, energy of desorption of ammonia (E) is calculated as an important parameter. E is written as follows,

$$E = E_{HZ} + E_{NH3} - E_{NH4Z}, \quad (4.1)$$

where E_{HZ}, E_{NH3}, and E_{NH4Z} are energies calculated for H-type zeolite, ammonia molecule, and H-type zeolite which accommodates ammonium cation, respectively. E is positive, and therefore $-E$ is energy for adsorption of ammonia on the Brønsted acid sites.

All the calculations stated here are performed on the DMol^3 software commercially available from Accelrys Co. Calculations are based on a generalized gradient approximation (GGA) level using Becke–Lee–Yang–Parr (BLYP) [2] or Hamprecht–Cohen–Tozer–Handy (HCTH) [3] exchange and correlation functional. Selection of the functional affects the calculation results strongly; GGA is recommended for the energy evaluation, and BLYP and HCTH are typical functionals working on the system [4]. Localized density approximation (LDA) is also used, but not recommended for use (*vide infra*). When the calculation is successfully completed, the geometry of the studied zeolite is optimized, and the minimum total energy for the optimized structure is calculated. Then E is given from (4.1) to be compared with the experimentally observed ΔU (change of internal energy upon desorption). The optimized geometry is used to justify the calculation because it can be directly compared with the experimental observation. The agreement between the calculated and experimental lattice parameters supports the theoretical calculation strongly. In addition, the precise geometry of the optimized structure of acid site affords us valuable information about the geometry and the strength of Brønsted acidity.

Various calculation methods are proposed, and precision and time cost depend on the method. The following is one of the examples selected in our calculations, and the calculated values are in satisfactorily agreement with the measurement values. Double numerical plus polarization (DNP) basis set is used for the numerical integration, and all the electrons are calculated. The convergence criteria of energy, force, and displacement are set to be of medium accuracy as 2×10^{-5} Ha, 4×10^{-3} Ha Å^{-1} and 0.005 Å (Ha: hartree, equal to 4.36×10^{-18} J), respectively.

4.1.2 Embedded Cluster and Periodic Boundary Conditions

A structure model is required in the initial step for the calculation. The commercially available software, for example Material Studio by Accelrys, is sold with an attached library package of crystal structure; therefore, the structure of usual zeolite is easily obtained. Web site of International Zeolite Association has a database of the structure in a crystal information file (CIF) format, from which the structure of zeolite is constructed for the calculation.

4.2 Application to Chabazite, a Simple Zeolite

Computational study requires time cost, which depends on the number of electron calculated. Therefore, very large models containing many elements are not practical for the calculation, but the precision of the calculation should be kept as high as possible. Cluster models are used for a fast and simple calculation. On the other hand, models prepared within the periodic boundary conditions are used for a more precise calculation. Examples of these models and calculation are shown below in detail.

4.2 Application to Chabazite, a Simple Zeolite

4.2.1 Brønsted Acid Sites in Chabazite Based on the Models Within the Periodic Boundary Conditions

Chabazite (CHA) has a simple structure with the characteristic that all T sites are equivalents. Therefore, only four kinds of Brønsted acid site are formed on the T sites with a tetrahedral configuration. This feature is found on the Y zeolite as well as on the CHA. The number of atoms in the unit cell of CHA is only 36, which is so small that a fast calculation is possible. CHA is therefore a zeolite to which the DFT is most easily applied with the structure model constructed within the periodic boundary conditions [5].

Figure 4.1 shows the structure of CHA within the periodic boundary conditions. Calculated geometrical parameters are compared with the experimentally observed values on H-SSZ-13, as shown in Table 4.1. H-SSZ-13 is a high silica zeolite with the structure of CHA. Calculated Si–O bond distances are 1.626–1.639 Å, and

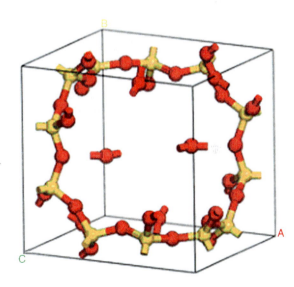

Fig. 4.1 Structure of chabazite (CHA) HAlSi$_{11}$O$_{24}$ constructed within the periodic boundary conditions

Table 4.1 Geometrical parameters of the optimized structures of siliceous CHA structure and experimental parameters of SSZ-13[a]

	Si-O (Å) Calc.	Si-O (Å) Exp.	SiOSi (°) Calc.	SiOSi (°) Exp.
O1	1.626	1.617	146.8	144.8
O2	1.630	1.613	151.0	150.0
O3	1.631	1.599	150.0	149.4
O4	1.639	1.615	147.6	147.8

[a]Experimental values are taken from the neutron diffraction study of the H-SSZ-13 (Si/Al$_2$ = 32) in [6]

Table 4.2 Comparison between energies of ammonia desorption on four kinds of Brønsted OH in chabazite calculated based on periodic boundary and embedded cluster models using different functionals, and experimentally observed ΔU

Acid sites	MR[a]	Periodic GGA-HCTH		Embedded cluster[b] GGA-BLYP	LDA-VWN	ΔU
O1H		128	127	139	204	131
O2H	8	129	125	137	207	134
O3H		131	129	139	207	128
O4H	6	110	111	127	204	101

[a]Oxygen-membered ring
[b]HAlSi$_{65}$O$_{102}$H$_{60}$

nearly the same as the experimental values 1.599–1.615 Å. Likewise, calculated Si–O–Si bond angles are 146.8–151.0°, and are very close to the experimental values 144.8–150.0°. Both parameters of distance and angle calculated are, therefore, in good agreement with the experimental observations; and the DFT calculation under the present conditions is proved to be a good method.

Table 4.2 shows the energies of ammonia desorption on the H-type CHA thus calculated. IR absorptions of OH bands have already been identified by a comparison with the data cited in literatures, and the energy of ammonia desorption has already been determined experimentally, as shown in Chap. 3. Because $\Delta H = \Delta U + P\Delta V$ and $\Delta V > 0$ for ammonia desorption, $\Delta U = \Delta H - RT$, where T is set to be the peak temperature. Thus corrected ΔU should be compared with the E calculated. As found from the table, values of the calculated E are divided into two energy regions, i.e., 128–131 and 110 kJ mol^{-1} for O1H, O2H, and O3H sites on the 8-ring, and O4H site on the double 6-rings, respectively. These calculated values are, therefore, approximately equal to those of experimentally observed ΔU. Thus, we can confirm that the heats of ammonia desorption experimentally measured by the IRMS-TPD are supported by the theoretical calculation. This coincidence is an example of the important progress for the characterization of the zeolite acidity, in which the DFT calculation is utilized to confirm the experimental observation.

4.2.2 Brønsted Acid Site in an Embedded Cluster Model

Periodic boundary conditions are not easily applicable because the calculation cost is very high. One method to overcome the problem is a utilization of a simple structure of an embedded cluster model. A Brønsted acid site cluster is embedded in a zeolite, and the energy is calculated totally on the zeolite with the cluster. The Brønsted acid site cluster is prepared from 8 T sites owned by seven Si and one Al, which is embedded in the CHA. The structure of the embedded Brønsted acid site cluster is optimized, but the surrounding siliceous CHA structure is not optimized, and the energy is calculated for the total structure with 66 T sites. Table 4.2 shows thus calculated energies for ammonia desorption based on the cluster model using various functionals. Energies obtained from the GGA-HCTH functional agree well with those obtained under the periodic boundary conditions using the HCTH functional. The calculation on the cluster model with the GGA-BLYP functional shows somewhat larger values and the LDA-VWN functional results in the energy values much larger than others. Therefore, the embedded cluster model is also applicable, as long as the BLYP or HCTH functional under the GGA level is utilized for the calculation.

From an application of DFT calculation to the acid site in CHA, we learn that the selection of either an LDA or a GGA potential affects the calculation strongly, and the GGA level is recommended. Furthermore, it is identified that the embedded cluster model is correctly working for the precise determination of the energy.

4.3 Application to Other Zeolites

4.3.1 FAU, MOR, and BEA Calculated Under the Conditions of the Embedded Cluster and the Periodic Boundary

DFT calculations are successfully applied to the CHA, a zeolite with the simple structure, based on the models prepared under the embedded cluster and the periodic boundary conditions, and the calculated energies for the ammonia desorption are satisfactorily agreed with the experimental values. In next step, therefore, our interests are directed to the application of these methods to other usual zeolites. However, application conditions for the calculation depend on the structure of zeolite; i.e., the time cost for the calculation depends strongly on the zeolite structure. Therefore, an application to FAU and MOR that have the relatively simple structure will be mentioned. In addition, because of an interest to the catalytic application, some of the Brønsted acid sites of BEA zeolite are also studied based on the same methodology.

As mentioned above, the FAU has a simple structure of zeolite, similarly to CHA, and all the T sites included are equivalents; therefore four kinds of OH are possibly located in the framework of Y zeolite. However, the O4H in the super cage is not detected by the neutron diffraction experiment. In other words, only three kinds of Brønsted OH are observed. Therefore, O1H in the super cage, O2H in the sodalite cage, and O3H inside the double 6-rings are calculated by the DFT method. Calculations are performed not only on the embedded cluster with BLYP and HCTH functionals [7], but also on the periodic boundary with HCTH functional [8], and these are compared with the measured ΔU. The embedded cluster model of FAU is shown in Fig. 4.2, in which a Brønsted acid site consisting of eight $T(\text{HAlSi}_7\text{O}_7)$ sites is included. Hydrogen is put on each terminal element, and total chemical formula in the calculated model of Y zeolite is $\text{HAlSi}_{47}\text{O}_{78}\text{H}_{37}$ (164 atoms). The precision of calculation depends not only on the number of atoms in the acid site cluster, but also on the size of surrounding zeolite groups. In the present calculation, two more outside neighbor T sites are calculated without optimization.

On the other hand, mordenite has four kinds of T sites and ten kinds of oxygen. Among four T sites, three (T_1, T_2, and T_4) are on the large pore of 12-ring, and one (T_3) on the small pore of 8-ring. Because of the stabilities and previous studies, eight kinds of Brønsted OH are selected, as shown in Table 4.3. Only the O2H is

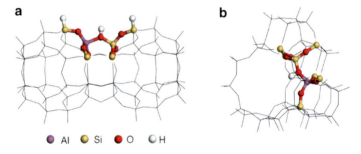

Fig. 4.2 Embedded cluster model of a Brønsted acid in FAU (**a**) and MFI (**b**)

Table 4.3 Comparison of the energies of ammonia desorption (kJ mol^{-1}) calculated on periodic boundary and embedded cluster models

			8 T cluster			Periodic		
Zeolite	Brønsted OH	MR	LDA-VWN	GGA-BLYP	GGA-HCTH	LDA-VWN	GGA-HCTH	ΔU^a
FAU	O1	12	177	100	100		122	105
	O2		196	110	101		110	115
	O3	6	178	93	86		101	101
MOR	Al1O3Si2		194	132	122	215	–	
	Al2O2Si4		211	143	131	213	–	142
	Al4O2Si2	12	206	140	129	–	–	
	Al2O5Si2		218	147	136	–	135	
	Al4O10Si4		210	143	129	220	141	
	Al3O1Si1		230	146	132	–	–	
	Al1O6Si1	8	239	155	142	–	136	147
	Al3O9Si3		238	142	126	235	150	
BEA	Al9O12Si4	12	194	133	119		127	125
	Al1O4Si8	6	197	119	116		122	

[a]Measured by the IRMS-TPD experiment

calculated in two different neighbor positions of Al, Al2, and Al4. The O6H in the Al1O6HSi1 is stabilized to be directed into the 8-ring, though it is attached to the Al1 in the 12-ring.

BEA has nine kinds of T sites and 17 kinds of oxygen. Among them, two sites are selected for the Brønsted OH, because ammonium cation is highly stabilized on these sites.

Table 4.3 shows the energies calculated for the embedded acid site cluster and the models prepared within the periodic boundary conditions on FAU, MOR, and BEA zeolites, in which various functionals are applied for a comparison. Energies calculated on the cluster models are compared; these values of energy are always in the sequence, LDA-VWN > GGA-BLYP > GGA-HCTH, and these values roughly correlate with each other. The energies calculated within the periodic boundary conditions using LDA-VWN are similar or a little larger than those obtained on the cluster models using the same functional. On the other hand, the energies obtained with the GGA-HCTH functional under the periodic conditions are similar to those obtained under the cluster models using the same functional. Therefore, the calculated values are influenced by the selected potential, LDA or GGA, but not so much by the selected functionals, similarly to the CHA as mentioned above. Because the calculated values should be compared with the experimentally observed values, the selection of GGA level is preferentially recommended, and BLYP and HCTH functionals based on periodic boundary and embedded cluster models are working well to afford us the reasonable values of ammonia desorption.

On the Y zeolite, it is found that the energy on the O3H in the 8-ring is smaller than that on others in the 12-ring, and the calculation is in agreement with the experimental observation. Furthermore, it is proved without exception that mordenite has the Brønsted acidity on the 8-ring which is stronger than that on the 12-ring. This calculation result on MOR is also supported by the experimental findings. In other words, the distribution of the Brønsted acid sites depending on the site location is consistently concluded based on any selected method of calculation, although the absolute values are somewhat different. Therefore, the experimental observations are strongly supported by the DFT calculation.

4.3.2 MFI, FER, and MWW Calculated Under the Embedded Cluster Model

Other zeolites contain various structurally different T sites, and therefore many kinds of the Brønsted acid sites are possibly stabilized, as shown in Table 4.4. The calculation is limited to the model of embedded cluster.

MFI has 12 kinds of T sites and 26 kinds of oxygen, and it is not easy to select the calculated Brønsted acid sites. Our calculation followed the selection by Simperler et al. [9] and five kinds of Brønsted acid sites are studied, as shown in Table 4.5. Likewise, four kinds of acid sites on FER and five kinds of acid sites on MWW are studied, as shown in Table 4.5.

Table 4.4 Number of different T and O sites and T site multiplicity

Zeolite code	Zeolite name	T sites	O sites	Multiplicity[a]
ANA	Analcime	1	1	48
LTA	Linde A	1	3	24
CHA	Chabazite	1	4	36
FAU	Y (Faujasite)	1	4	192
FER	Ferrierite	4	8	36
MOR	Mordenite	4	10	48
MEL	ZSM-11	7	15	96
MWW	MCM-22	8	13	72
BEA	β	9	17	64
MFI	ZSM-5	12	26	96

[a]The number of T sites in a unit cell

Table 4.5 Energies of ammonia desorption (kJ mol^{-1}) calculated on embedded cluster models

Zeolite	Brønsted OH	MR	8 T cluster LDA-VWN	GGA-BLYP	GGA-HCTH	ΔU
	Al11O11Si12		198	137	126	132
	Al7O17Si4		213	146	136	
MFI	Al12O24Si12	10	189	125	117	
	Al7O7Si8		202	142	131	
	Al9O18Si6		196	135	125	
FER	Al2O7Si4	10	202	123	111	135
	Al4O6Si4	10	198	124	114	
	Al3O4Si1	Cage	204	135	126	
	Al2O1Si2	10	209	132	127	
MWW	Al4O3Si1	12	213	156	142	
	Al3O1Si2	10	196	134	128	
	Al5O9Si2	10	203	140	124	
	Al5O6Si4	12	208	142	127	135
	Al5O8Si5	12	204	140	127	
	Al6OSi1	10	201	133	119	

As found from the table, a similar trend of the calculated values is observed; i.e., the energies are in the sequence, on LDA-VWN > on GGA-BLYP > on GGA-HCTH. A comparison between the calculated energies and the experimental ΔU does not lead to a simple conclusion of the consistence. So far, the location of Al is not known yet, and therefore, a strict comparison of the Brønsted OH is difficult. We can understand from the calculation, however, the upper and lower limits of the energy values, and the experimental value of ΔU is within this region.

Finally, a comparison between the calculated values of E and the experimentally observed ΔU is summarized in Fig. 4.3. The values obtained using the GGA-BLYP functional on the embedded cluster model and using the GGA-HCTH functional under the periodic boundary conditions are compared with the experimentally

4.4 Modified Zeolites

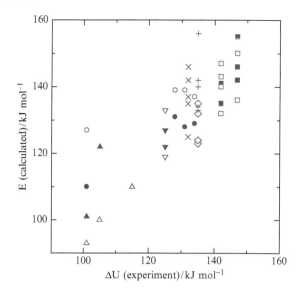

Fig. 4.3 Comparison between E calculated and ΔU experimentally measured on zeolites, CHA (*filled circle, open circle*), FAU (*filled triangle, open triangle*), MOR (*filled square, open square*), BEA (*filled inverted triangle, open inverted triangle*), FER (*open diamond*), MWW (*positive symbol*), and MFI (*multiplication symbol*). Energies were calculated under the periodic boundary conditions with the HCTH functional (*closed symbols*) and the embedded cluster models with the BLYP functional (*open symbols, multiplication symbol* and *positive symbol*)

observed values. It is found that the periodic boundary conditions provide the value which is more consistent with the experimental observation than the cluster model. Large deviations are found on the comparison in MFI and MWW. However, maybe, the scattered values could be neglected because the locations of acid sites are not found exactly, and the acid site on a scattered location may not be stabilized energetically.

4.4 Modified Zeolites

4.4.1 Divalent Cation-Exchanged Y Zeolites Based on the Embedded Cluster Model

Computational calculations as shown above support the experimental observation of IRMS-TPD. Although the absolute values of energy depend on the method of calculation, the calculated energy values tell us the sequence of zeolites which enables us to understand their acid strengths. This method of DFT calculation, therefore, can be applied to the modified zeolites, and the change of solid acidity and catalytic activity caused by the modification is studied in detail. Cation-exchanged Y zeolites are now studied by the DFT calculations [10].

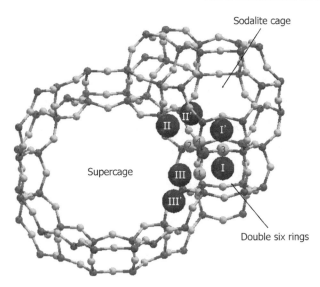

Fig. 4.4 Structure of Y zeolite with definitions of the oxygen sites 1, 2, 3, and 4 and the cation-exchanged sites I, I′, II, II′, III, and III′

As mentioned in the last chapter, Ca, Ba, and La ion-exchanged Y zeolites have been utilized for the catalytic reaction, and the enhanced acidity is an interesting subject of the catalytic chemistry. Physical chemistry observed in the experiment is now thoroughly studied from the view-point of the computational calculation.

The IRMS-TPD experiment on these cation-exchanged Y zeolites reveals the enhancement of the Brønsted acidity. Therefore, calculations are performed on the Brønsted OH in the Y zeolite, and the mechanism of the enhancement of acid site strength is studied. First, the structure of the cation-exchanged Y zeolite is created. Previous studies of inorganic chemistry determined the site of alkaline earth cation exchanged in a Y zeolite, and sites of I, I′, II, II′, and III are selected for the divalent cation $M = Mg^{2+}$, Ca^{2+}, and Ba^{2+}, as shown in Fig. 4.4. Site occupancy depends on the cation, and these cations are preferentially located on the sites I, I′, II, and II′. These sites are located close to the sodalite cage, and have a chance of interacting with the OH in the sodalite cage. The position of cations' location is in agreement with the infrared observation of OH, because the OH in the sodalite, O2H and O3H, diminishes preferentially by the exchange of cations (Fig. 3.11). The Brønsted OH sites in the super cage still remain after the ion exchange, and the strength of one of the OH bands is enhanced according to the experimental observation of IRMS-TPD. DFT calculations are studied on the embedded cluster model modified by the exchange of cation.

Table 4.6 shows the calculated heats of ammonia desorption on the O1H in the super cage. LaOH cation is also assumed to be located, and studied simultaneously. When a comparison is made between the calculated and observed values of energies, cations located in sites I′ and II are most probable because the Brønsted OH is enhanced by the cation exchange on these sites. Metal cations are regarded as electron acceptors; thus, the Brønsted OH situated close to the cation becomes elec-

4.4 Modified Zeolites

Table 4.6 Energy of ammonia desorption on the O1H unmodified and modified by the divalent cation

Metal cation	E (DFT) (kJ mol^{-1})			ΔU (TPD) (kJ mol^{-1})
	Site I	Site I'	Site II	
(no)	99	100	95	106
Ba	116	104	109	114
Ca	97	106	116	118
LaOH	–	110	–	114

tron deficient, thus becoming a strong Brønsted acid site. Increment of the acid site strength introduced by the divalent cation is thus theoretically supported.

4.4.2 Modified Brønsted OH in Y Zeolite Based on the Periodic Boundary Conditions

In the last chapter, La cation is studied with an assumption that La cation is stabilized as the divalent La-OH^{2+}. Because of the basicity, lanthanum cation readily becomes stabilized as a cation containing hydroxyl group. Similarly, Al cation in the extra framework of Y zeolite may be stabilized as Al-OH^{2+} in the cation-exchanged sites, thus enhancing the Brønsted acidity. The metal hydroxide formation is known because of the equation,

$$M^{2+} + H_2O + Si - O - Al \rightarrow M(OH)^+ + Si - (OH^+) - Al. \quad (4.2)$$

This reaction takes place readily on the metal cation with a high electrostatic potential (= valence/cation size). Al^{3+} has a high electrostatic potential (5.61 Å^{-1}); thus it is readily hydrolyzed. Al atom in the framework is released during the steaming of Y zeolite at high temperatures to produce the ultra stable Y (USY) zeolite, and is stabilized in exchange sites of Y zeolite. Therefore, DFT calculations are studied under the periodic boundary conditions in order to understand the production of active Brønsted OH in USY zeolite in more detail [11].

Table 4.7 shows the list of the calculated energy for divalent cation-exchanged Y zeolites under the periodic boundary conditions. Ba and Ca are assumed to be located in the cation-exchanged sites, I, I', and II, while Al-OH is in the sites I' and II'. The Brønsted O1H on unmodified H-Y zeolite is calculated under the environment of 1Al or 3Al; however, nearly the same value of E is calculated. Enhancement of the Brønsted acid strength is calculated, and as shown in Table 4.7, the values of E calculated on the conditions of metal cations located on I', II, and II' are reasonably agreed with the experimental observation. Thus, the present calculation also supports the IRMS-TPD experiment, and the exchange sites I', II, and II' are probably occupied by these cations.

Table 4.7 Energy of ammonia desorption and IR absorption band shift of the O1H unmodified and modified by the divalent cation

	Cation sites	EN[a] of added element	E (kJ mol^{-1}) Calculated (DFT)	ΔU (kJ mol^{-1}) Measured (TPD)	$\Delta \nu$ (cm^{-1}) Calculated (DFT)	$\Delta \nu$ (cm^{-1}) Measured (TPD)
H$^+$ (1Al)			122	105	(3,797)	(3,648)
H$^+$ (3Al)			119			
H$^+$Ba^{2+} (3Al)	I	0.78	118	114	−6	−16
	I′		135		−21	
	II		122		−10	
H$^+$Ca^{2+} (3Al)	I		122		5	
	I′	1.22	136	118	−36	−16
	II		128		−14	
H$^+$AlOH^{2+} (3Al)	I′	3.47	144	133	−48	−53
	II′		135		−46	

[a]Electro-negativity

The stretching vibration of OH is also studied for these cation-exchanged Y zeolites. Because of the assumption of anharmonic oscillation, the absolute IR absorption band calculated on the unmodified H-Y zeolite is 3,797 cm^{-1}, a little larger than the experimental value of 3,648 cm^{-1}. Therefore, only the shift of IR band position ($\Delta \nu$) due to the introduction of metal cation is compared with the experimental finding. As shown in Table 4.7, the stretching vibration of O1H shifts to lower wavenumber to confirm the experimental observation. Therefore, the IR observation is also supported by DFT calculation.

The calculated model of the USY, in which Al-OH is located in the sites I′ in the sodalite cage, is shown in Fig. 4.5. Extra-framework Al is surrounded by three oxygen species in the zeolite wall and one more OH; bond lengths of three Al–O and one Al–OH are 1.86–1.94 and 1.71 Å, respectively, and therefore the structure could be regarded as the pseudo tetrahedron. The structure is in agreement with suspects from ^{27}Al MQMAS NMR study [12].

4.5 Dependence of Brønsted Acid Strength on Local Geometry [13]

In Sects. 4.1–4.3, the DFT calculation is applied to typical zeolites, and the optimized geometry of Brønsted acid sites Si(OH)Al and the energy of ammonia desorption are provided. Over a wide range of zeolites, DFT calculations and ammonia IRMS-TPD experiments give a reasonable agreement between the ammonia adsorption energies calculated and measured. This means that the clusters or periodic unit cells assumed in such DFT calculations are reliable from a view of their

4.5 Dependence of Brønsted Acid Strength on Local Geometry

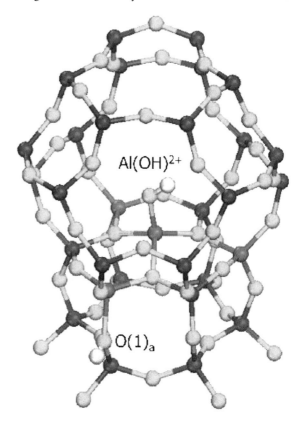

Fig. 4.5 The optimized structure model of USY; Al-OH is located in the site I' in the sodalite cage

structures. Thus, the database of the acid sites about the structure and energy is created. Therefore, based on the calculated parameters, it may be possible to find how the acid strength is controlled by the framework structure (topology).

It is presumed that the local geometry, i.e., bond lengths and angles among atoms surrounding the acid site, controls the acid strength through some quantum chemical effects. The correlation between the geometric and energy parameters will help us to design a new type zeolite with a desired strong acidity. Purely theoretical studies have proposed that the shorter Al–O bond and the larger Si–O–Al angle bring about the stronger acidity [14,15]. Our study is devoted to find the parameter which affects the acid strength most strongly through the experimental confirmation, and then to correlate these quantum effects with the crystal topology of the strong acid site.

The structural parameters shown in Fig. 4.6 are calculated, and correlations between E (ammonia desorption energy) and these geometrical parameters are analyzed. Relationships among the parameters are analyzed based on the following assumptions.

1. The acid strength is controlled by the local structure of the Si(OH)Al unit. The local structure can be expressed by parameters such as Al–O_a distance (a) and

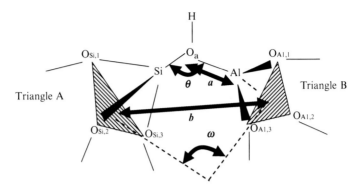

Fig. 4.6 Geometric parameters. *b* is the distance between centers of triangles A [$O_{Si,1}$-$O_{Si,2}$-$O_{Si,3}$] and B [$O_{Al,1}$-$O_{Al,2}$-$O_{Al,3}$]. ω is the planar angle between the two triangles

Si–O_a–Al angle (θ). Because electrons in an SiOH group can be withdrawn by the Lewis acidic Al to cause the Brønsted acidic property, the shorter the *a*, the less negative the O_aH charge, resulting in the higher acid strength. On the other hand, the *p*-character of the O_a–H bond should be increased with increasing θ. The increase in *p*-character should correspond to weakness of the O_a–H bond, and therefore, larger θ should render the higher acid strength [14, 15].

2. The crystal structure determines the positions of atoms surrounding the Si(OH)Al unit, and these positions decide the local structure. In order to express mathematically the influence of the positions of surrounding atoms, the parameters *b* and ω can be used, because the atoms forming triangles A and B (Fig. 4.6) are close to the outer part fixed as the crystal structure.

Therefore, relationships among *E*, *a*, and θ are first analyzed to clarify (1) whether a negative relationship between *a* and *E* exists; (2) whether a positive relationship between θ and *E* exists. Then, relationships among *a*, θ, *b*, and ω are analyzed to show whether the relations can be correlated to acid strength.

Figure 4.7 shows that a simple relationship is not obtained between *a* and *E*. This is because the coordination environment of NH_4^+ affects the desorption heat. Formation of the NH_4^+ species B and C shown in Fig. 4.8, which have exceptionally short distances to the wall oxygen species, is probably affected by the confinement effect [16] or the steric hindrance. In contrast, the species A is believed to show the ammonia desorption energy reflecting the intrinsic acidity. In Fig. 4.7, the group A is shown by filled symbols. At least with respect to A, a relationship between *E* and *a* can be drawn, with the shorter *a* giving the higher acid strength (higher *E*). In contrast, any dependence of *E* is neither found on θ nor on φ.

A more in-depth rationalization of the reason that explains the correlation between *a* and *E* can be seen in Fig. 4.9. An inverse relationship between O_aH Mulliken charge and *a* shown here suggests again that the intrinsic origin of the Brønsted acidic property is due to the electron withdrawing from O_aH by the Lewis

4.5 Dependence of Brønsted Acid Strength on Local Geometry

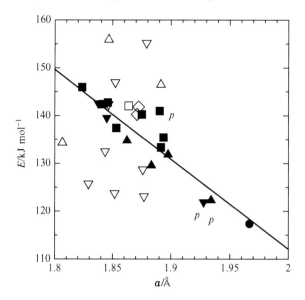

Fig. 4.7 Plots of E (calculated value) against a (Al–O_a distance). The coordination environment of NH_4^+ is shown by the following symbols; the coordination environment is A (*filled circle, filled triangle, filled square, and filled inverted triangle*: H = H−, H = H−*, H = H = *, and H − H−, respectively), B (*open triangle and open inverted triangle* for H ≡ H−* and H ≡ H = *, respectively) or C (*open square* and *open diamond* for H ≡ H = H = ** and H − H − H−, respectively) species shown in Fig. 4.8. p shows the results based on the periodic boundary conditions, while the others are based on the cluster method

acidic Al. The shorter a yields the stronger interaction between Al and O_aH; thus, it results in the smaller charge and the higher acid strength of O_aH.

In Fig. 4.8, we notice a following relationship between E and a:

$$E \approx 488 - 187a, \tag{4.3}$$

where E and a are shown in kJ mol^{-1} and Å, respectively. On the other hand, a numerical analysis shows that the distance a can be expressed by a simple function of b and ω as follows.

$$a = 0.29b + 0.0037\omega + 0.56. \tag{4.4}$$

This supports that the relationship is purely geometric; (1) a is strained when b and/or ω are increased; (2) longer b gives longer lengths for all bonds in the Si(OH)Al unit when ω is fixed; (3) larger ω should move O_a to the upper direction shown in Fig. 4.6, elongating a when b is fixed.

By combining these relationships, a simple relationship is found, as shown in Fig. 4.10.

$$E \approx 419 - (67b + 0.84\omega), \tag{4.5}$$

where E, b, and ω are shown in kJ mol^{-1}, Å, and degree, respectively.

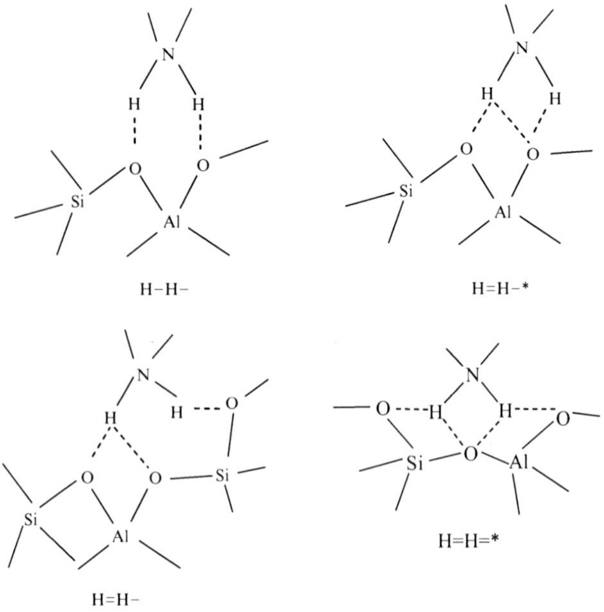

A: Bidentates in relatively open space.

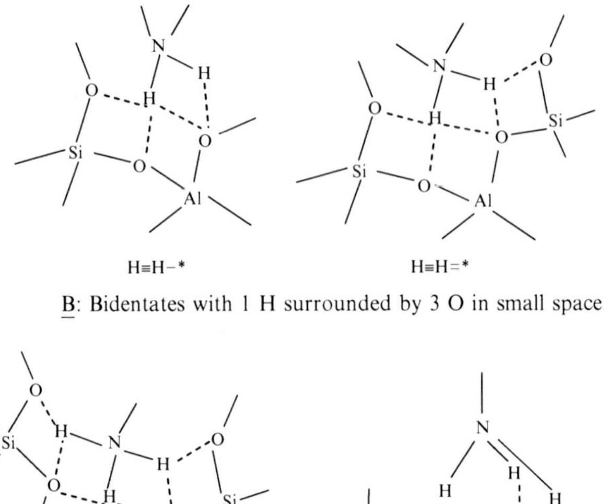

B: Bidentates with 1 H surrounded by 3 O in small space.

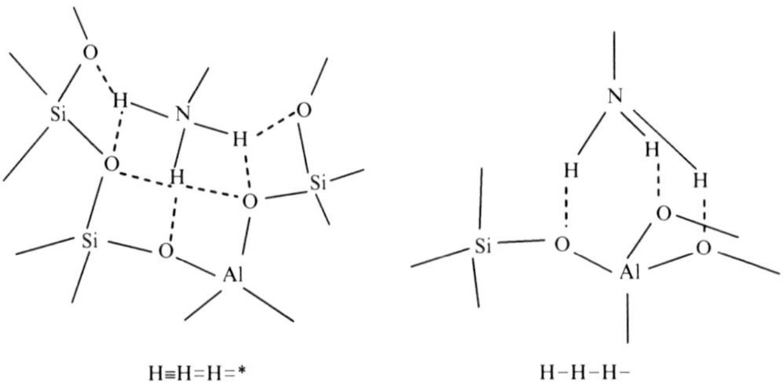

C: Tridentates.

Fig. 4.8 Schematic drawings of coordination environments of NH_4–zeolite. Dotted lines between O and H show short interatomic distances less than 2.7 Å

4.5 Dependence of Brønsted Acid Strength on Local Geometry

Fig. 4.9 Relationship between Mulliken charge e_{OaH} and a (Al–O_a distance). *Symbols* show the crystal type and coordination environment as defined in Fig. 4.7

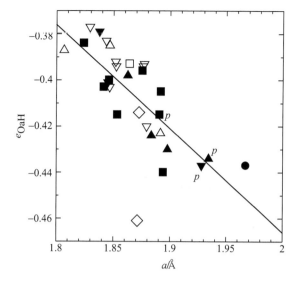

Fig. 4.10 Dependence of E_{ads} (calculated value) on $58b + 0.79\omega$. *Symbols* show the crystal type and coordination environment as defined in Fig. 4.7

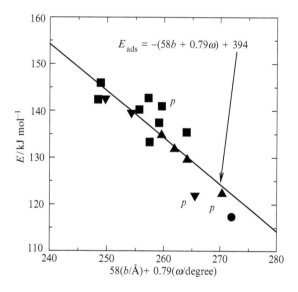

This equation gives us an important conclusion that the topological centers with small b and ω give the higher Brønsted acid strength. When the Si(OH)Al unit is pushed from both sides, the acid strength becomes higher, because the Al–O_a distance is shortened. Zeolite framework topology that realizes such a situation is the origin of acid strength.

Interestingly, the strong acid sites have a trend to be found in relatively small micropores. This is observed in the acid strength order of MFI > BEA > FAU, 8-ring > 12-ring in MOR or 6-ring > 12-ring in FAU, as shown in Chaps. 2 and 3. The present finding also explains that it has been difficult to detect strong acid sites on the so-called mesoporous zeolites, although many attempts have been carried out. On a flat surface or on a wall of relatively wide pore with a small curvature, b and ω should be large, resulting in the low acid strength according to the present proposal. From these considerations, an important conclusion can be extracted; a strong acid site is difficult to be constructed on an opened space. This is very important information for a design of new catalysts especially for the reaction of heavy hydrocarbons that is demanded to establish an alternative fuel technology.

Nevertheless, there remain some possibilities. Because the zeolite framework is complex, it is possible to realize both the strong acidity and the high accessibility by a precise design of the structure based on this information. An easier strategy is to utilize the extra-framework cation because the extra-framework Al or other cations can vary the acid strength based on different rules, as stated in the previous sections.

References

1. W. Kohn, http://nobelprize.org/nobel_prizes/chemistry/laureates/1998/kohn-lecture.pdf
2. A.D. Becke, J. Chem. Phys. **104**, 1040 (1996)
3. F.A. Hamprecht, A.J. Cohen, D.J. Tozer, N.C. Handy, J. Chem. Phys. **109**, 6264 (1998)
4. A.D. Corso, A. Pasquarello, A. Balderesch, R. Car, Phys. Rev. B **53**, 1180 (1996)
5. K. Suzuki, G. Sastre, N. Katada, M. Niwa, Phys. Chem. Chem. Phys. **9**, 5980 (2007)
6. L.J. Smith, A. Davidson, A.K. Cheetham, Catal. Lett. **49**, 143 (1997)
7. K. Suzuki, G. Sastre, N. Katada, M. Niwa, Chem. Lett. **36**, 1034 (2007)
8. K. Suzuki, G. Sastre, N. Katada, M. Niwa, Chem. Lett. **38**, 354 (2009)
9. A. Simperler, R.G. Bell, M.D. Foster, A.E. Gray, D.W. Lewis, M.A. Anderson, J. Phys. Chem. B **108**, 7152 (2004)
10. T. Noda, K. Suzuki, N. Katada, M. Niwa, J. Catal. **259**, 203 (2008)
11. K. Suzuki, T. Noda, G. Sastre, N. Katada, M. Niwa, J. Phys. Chem. C **113**, 5672 (2009)
12. N. Katada, S. Nakata, S. Kato, K. Kanehashi, K. Saito, M. Niwa, J. Mol. Catal. A Chem. **236**, 239 (2005)
13. N. Katada, K. Suzuki, T. Noda, G. Sastre, M. Niwa, J. Phys. Chem. C **113**, 19208 (2009)
14. H. Kawakami, S. Yoshida, T. Yonezawa, J. Chem. Soc. Faraday Trans. **80**(2), 205 (1984)
15. I.N. Senchenya, V.B. Kazansky, S. Beran, J. Phys. Chem. **90**, 4857 (1986)
16. G. Sastre, A. Corma, J. Mol. Catal. A Chem. **305**, 3 (2009)

Chapter 5
Catalytic Activity and Adsorption Property

Abstract Functions of zeolites as catalysts and adsorbents are related with the observed acidic properties. The turnover frequency and activation energy of alkane cracking depend on the ammonia desorption heat. Desorption heat of toluene on Na-zeolite is related with the ammonia adsorption heat of corresponding H-zeolite. Highly active catalysts for Friedel–Crafts alkylation (Ga/MCM-41) and amination of phenol (Ga/ZSM-5) are found based on the ammonia TPD analysis.

5.1 Paraffin Cracking

5.1.1 Evaluation of Intrinsic Activity of Acid Site

Cracking of alkane (paraffin) allows mankind to convert petroleum into gasoline and other necessary feedstocks efficiently. Natural clay materials, synthesized amorphous silica-aluminas, and, at last, USY (ultrastable Y) zeolite-based materials have been used as main components of the catalysts for fluid catalytic cracking (FCC) process since 1960s. Subsequently, so-called high silica zeolites, e.g., ZSM-5, β and MCM-22, have been found to possess high activities for the alkane cracking, and they are used as catalysts or catalyst additives for specific purposes. Solid acidity of these catalysts is the intrinsic origin of the catalytic activity. In this section, quantitative analysis of the role of acid site on the catalytic cracking of alkane is reviewed.

According to Haag and Dessau, a penta-coordinated carbonium cation (**1**) is formed from an alkane and a Brønsted acid site on a zeolite as (5.1) [1].

$$R-CH_2-CH_2-CH_2-R' + H^+ \rightarrow R-\underset{\underset{H\quad H}{\diagup\diagdown}}{\overset{\overset{H}{|}}{C^+}}-CH_2-CH_2-R' \qquad (5.1)$$

1

Later, the microstructure of **1** is drawn as **2** or **3**, or it is interpreted that an equilibrium (5.2) exists between the isomers [2–4]. All these isomers are so-called nonclassic carbonium cations that are formed by the protonation of alkane.

$$\underset{2}{R-\underset{H}{\overset{H}{\underset{|}{C}}}-H^{+}-CH_{2}-CH_{2}-R'} \rightleftarrows \underset{3}{R-\underset{H\ \ H}{\overset{H}{\underset{|}{C}}}\cdots\overset{+}{\cdots}\cdots CH_{2}-CH_{2}-R'} \rightleftarrows \underset{1}{R-\underset{H\ \ H}{\overset{H}{\underset{|+}{C}}}-CH_{2}-CH_{2}-R'} \quad (5.2)$$

Following the formation of such a carbonium cation, subsequent reactions readily proceed. For example, a reaction (5.3), cracking of the carbonium cation into a short alkane, and a carbenium cation **4** is possible.

$$R-\underset{H\ \ H}{\overset{H}{\underset{|+}{C}}}\cdots\cdots CH_{2}-CH_{2}-R' \rightarrow R-CH_{3} + \underset{4}{H_{2}C^{+}-CH_{2}-R} \quad (5.3)$$

The carbenium cation, which has a triangle sp^2 hybrid orbital and a void π orbital, readily reacts via various pathways. When a proton in the carbenium cation is rebounded to the Brønsted acid site, an alkene (olefin) molecule is formed by (5.4).

$$H_2C^+\ CH_2-R \rightarrow H_2C = CH-R + H^+ \quad (5.4)$$

Through reactions (5.1), (5.3), and (5.4), a long alkane is cracked into a pair of a short alkane and a short alkene. From the carbenium cation, other reactions such as alkylation, isomerization, dimerization, and further cracking (5.5) are also possible.

$$R-HC = CH-CH_2-R' + H^+ \rightarrow R-HC^+-CH_2-CH_2-R' \rightarrow R-HC$$
$$= CH_2 + C^+H_2-R' \quad (5.5)$$

Aromatic compound also forms a carbenium cation, and various reactions are possible.

Haag and Dessau have clarified the following reaction mechanisms through careful analysis of kinetics [1].

- At relatively low temperatures, high conversions, and/or high partial pressures of reactants, the successive reactions of the formed carbenium cation are predominant under bimolecular mechanism. The carbenium cation plays an important role to determine the total reaction rate under these conditions.

5.1 Paraffin Cracking

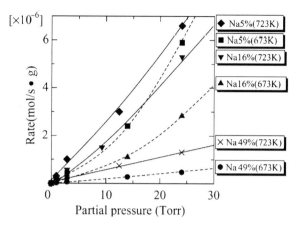

Fig. 5.1 Reaction rate of octane cracking on NaH-ZSM-5 zeolite as a function of partial pressure of octane. Reproduced with permission from [5]. Copyright Elsevier 2004

- At relatively high temperatures, low conversions, and/or low partial pressures of reactants, the rate-determining step is the formation of nonclassical carbonium cation **1** under monomolecular mechanism, and thereby the number and strength of Brønsted acid site determine the total reaction rate.

In the former mechanism, the reaction between two intermediates is important, and therefore the reaction rate should neither be linearly proportional to the partial pressure of reactant nor the number of active site. In addition, the extraction of a hydride anion from an alkane by Lewis acid site can form a carbenium cation, which possibly contributes to the total reaction rate. The contribution of Brønsted acid site should be unclear under the conditions where the bimolecular mechanism is predominant. On the contrary, the intrinsic contribution of active site is evaluated under the latter mechanism.

As shown in Fig. 5.1, the reaction rate of octane cracking shows the first-order dependency on the partial pressure of octane at 723 K, while the reaction order is high at 673 K. This evidences that monomolecular mechanism proceeds above 723 K. Under such conditions (723 K), the reaction rate on NaH-ZSM-5 zeolite shows the first-order dependency also on the number of Brønsted acid site (Fig. 5.2), which is measured by the ammonia IRMS-TPD method with varying the Na content [5]. The turnover frequency (TOF) can be determined from the slope, and it shows the intrinsic contribution of Brønsted acid site to the catalytic cracking. It should be noted that the intrinsic TOF cannot be obtained at lower temperatures where the bimolecular mechanism affects the total reaction rate.

5.1.2 Dependence of Activity on Acid Strength

Relationships between the TOFs for hexane and octane cracking at 773 K and the Brønsted acid strength (ammonia desorption heat) under monomolecular conditions are shown in Fig. 5.3a, b. At $< 135\,\text{kJ}\,\text{mol}^{-1}$ of the ammonia desorption heat, the

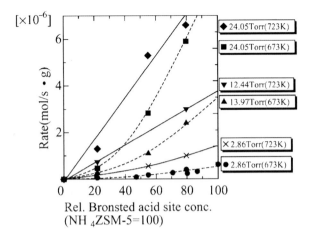

Fig. 5.2 Reaction rate of octane cracking as a function of number of Brønsted acid sites. Reproduced with permission from [5]. Copyright Elsevier 2004

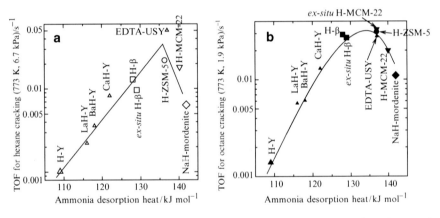

Fig. 5.3 Relationship between TOFs for cracking of hexane (**a**) and octane (**b**) and ammonia desorption heat of Brønsted acid site

TOF increases with increasing the acid strength. In this region, the logarithm of TOF shows a linear relationship with the ammonia desorption heat. The activation energy measured around 773 K also shows a linear relationship with the ammonia desorption heat at $< 140\,\text{kJ mol}^{-1}$, as shown in Fig. 5.4 [6]. These relationships clearly indicate the contribution of acidic property of zeolite to the catalytic activity for the reaction of alkane. The Brønsted acid site catalyzes the monomolecular mechanism of alkane cracking, and the stronger Brønsted acid brings about the higher reaction rate via the reduction of activation energy.

Such a simple explanation was not obtained previously, and the interpretation of cracking activity of zeolite has been under controversy. The main reason is caused by the difficulty in the measurements of solid acidity, especially on the USY zeolite,

5.1 Paraffin Cracking

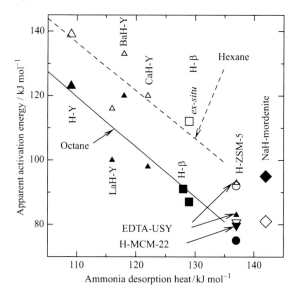

Fig. 5.4 Plots of activation energies for cracking of hexane (*white*) and octane (*black*) against ammonia desorption heat experimentally observed on H-Y (*filled triangle and open triangle*), H-β (*filled square and open square*), H-ZSM-5 (*filled circle and open circle*), ex situ H-MCM-22 (*filled inverted triangle and open inverted triangle*), NaH-mordenite (*filled diamond and open diamond*), and cation-exchanged Y zeolites (*filled small triangle and open small triangle*)

which is the most important component of cracking catalyst. It is invoked in a literature that at least 17 kJ mol^{-1} of the difference in ammonia desorption heats is needed to explain the large difference in the catalytic activities between H-Y and USY [7]; however, previous measurements of ammonia desorption heat by means of the ammonia TPD and microcalorimetry could not detect such a large difference between H-Y and USY. The readers of this book have already known that the ammonia IRMS-TPD shows a clear difference in the ammonia desorption heats (ca. 20 kJ mol^{-1}) between H-Y and USY, as presented in Chaps. 2–4. Previous studies without this knowledge proposed other interpretations as follows.

- The apparent activity of catalyst depends on the diffusion rate and hence on the mesoporosity. Especially, the enhancement of activity by steaming of Y into USY is mainly owing to the formation of mesopores [8].
- As shown later, physical adsorption heat of an alkane on zeolites apparently compensates the activation energy of cracking of the alkane. The activation step seems to be controlled by the physical adsorption property [9].

Steaming of an NH$_4$-Y zeolite at such a temperature as 823 K forms a USY zeolite with a high mesopore capacity, and simultaneously generates a high activity. The reactivity of bulky molecules such as alkyl aromatics and branched alkanes may be affected by the accessible surface area including the wall of mesopore. However, this is questionable at least for linear alkanes. Treatment of USY with Na$_2$H$_2$-EDTA solution generates strong Brønsted acid sites, because the extra-framework Al(OH)$^{2+}$

in the sodalite cage is increased by this treatment to enhance the Brønsted acid site in 12-ring, as shown in Chaps. 3 and 4. The change in activity is in good agreement with the number of strong Brønsted acid site, but not with the mesopore volume [10]. Moreover, TOF is quantitatively correlated with the strength of Brønsted acid site as shown above.

It has been reported that the physical adsorption heat of alkane (more exactly, the desorption enthalpy of physically adsorbed alkane molecule) is in the order of H-ZSM-5 > H-β > H-Y [9, 11], which is approximately dependent on the framework density of zeolite [12]. This is also a reverse sequence of micropore size, indicating that the physical adsorption property of alkane is mainly controlled by the curvature of micropore wall, which is believed to induce the confinement effect [13]. The order of cracking activity is also H-ZSM-5 > H-β > H-Y. Thus, it looks as if the physical adsorption property of zeolite, namely the confinement effect, controls the activity.

A broader study, however, makes it clear that the activity is not completely related with the physical adsorption heat. In Fig. 5.5, the activation energies of alkane cracking are plotted against the physical adsorption heats of corresponding alkane. One can find an inverse and linear relationship between the activation energy and physical adsorption heat over H-Y (**1**), H-β (**5**), and H-ZSM-5 (**6**), as shown by a dotted line in Fig. 5.5. It is noteworthy that the dependence of activity on the physical adsorption property was proposed from the results mainly on these zeolites. When other zeolites such as ion-exchanged Y and MCM-22 are taken into account, there is no relationship between the activation energy and adsorption heat. The

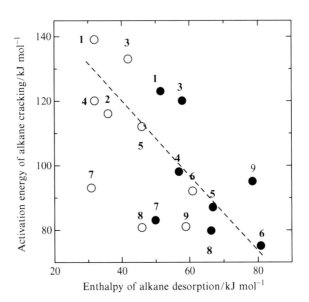

Fig. 5.5 Plots of activation energies for cracking of hexane (*open*) and octane (*filled*) against adsorption heats (desorption enthalpies) of hexane and octane, respectively, on H-Y (**1**), LaH-Y (**2**), BaH-Y (**3**), CaH-Y (**4**), ex situ H-β (**5**), H-ZSM-5 (**6**), EDTA-USY (**7**), H-MCM-22 (**8**), and NaH-mordenite (**9**)

5.1 Paraffin Cracking

most important finding is observed on the EDTA-USY (**7** in Fig. 5.5). The heat of alkane adsorption of EDTA-USY is close to that on H-Y, as well as those on CaH-Y, BaH-Y, and LaH-Y. This tells us that the physical adsorption property is constant when the framework topology is fixed, but the cation is changed. The activation energy on EDTA-USY is much lower than that on H-Y. This should be ascribed to the strong Brønsted acidity of EDTA-USY shown in Chaps. 3 and 4. Also on H-MCM-22, the activation energy is low although the alkane adsorption heat is low. The activation energy depends on the Brønsted acid strength, but not on the physical adsorption property.

Then, why does it look as if the activation energy depends on the physical adsorption property? On most of H-form zeolites such as H-Y, H-β, and H-ZSM-5, the desorption enthalpy of alkane and the ammonia desorption heat (acid strength) can be correlated, as shown in Fig. 5.6. The higher the desorption enthalpy of alkane, the higher the ammonia desorption heat on these zeolites. This is because the acid strength is controlled by the strain around the acid site, as shown in Chap. 4. The stronger compression from both sides of a SiOHAl unit gives the stronger acidity to this group. The compression becomes strong usually in a narrow micropore wall with a high curvature. On the other hand, the stronger physical adsorption property is due to the high curvature of micropore wall. Therefore, both the strong acidity and the high physical adsorption heat of alkane are observed in an associate manner on these zeolites. However, exceptions are possible for the curvature–acid strength relationship. On the ion-exchanged Y zeolites, the acid strength is controlled by a completely different manner, namely, the electron withdrawing nature of the extra-framework multivalent cation, as shown in Chap. 4. In addition, without such

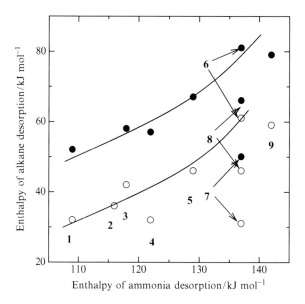

Fig. 5.6 Plots of physical adsorption heats (desorption enthalpies) of hexane (*open*) and octane (*filled*) against desorption heat of ammonia on H-Y (**1**), LaH-Y (**2**), BaH-Y (**3**), CaH-Y (**4**), ex situ H-β (**5**), H-ZSM-5 (**6**), EDTA-USY (**7**), H-MCM-22 (**8**), and NaH-mordenite (**9**)

extra-framework cations, there can be exceptions where the relationship between the curvature and acid strength is ruled out, because the zeolite framework structure is complex. H-MCM-22 seems to be the case where the acid strength is high although it has a relatively large micropore. Therefore, the acid strength is higher than on H-β, whereas the physical adsorption heat is similar. In these exceptional cases, the activation energy depends on the acid strength but not on the physical adsorption heat at all. The advancement in the measurement method of the acidic property of solid thus provides more precise interpretation [6].

5.1.3 Thermodynamic Description on Correlation Between Activation Energy and Ammonia Desorption Heat

The following discussion further clarifies the meaning of the relationship between ammonia desorption heat and activation energy.

It is believed that pK_as of conjugated acids of a series of basic compounds have a parallel relationship in different acidic solvents as (5.6) or (5.7) [14]. For any acidic solvent AH,

$$pK_a\left(B_iH^+/AH\right) - pK_a\left(B_{ii}H^+/AH\right) = \text{const}, \tag{5.6}$$

$$\frac{K_a\left(B_iH^+/AH\right)}{K_a\left(B_{ii}H^+/AH\right)} = \text{const}, \tag{5.7}$$

where B_i and B_{ii} are two kinds of bases, AH is an acidic solvent, BH^+ is the conjugated acid of a base B, and $pK_a\left(BH^+/AH\right)$ is the pK_a of BH^+ in AH.[1] Although these relationships have been confirmed only in the cases where acetic acid and water were the acidic solvents and amines were the bases [14], traditional theories on acid–base reactions such as the definition of H_0 index [15] are based on the assumed equation (5.6), which holds for the acids and bases discussed without exception.

On a Brønsted acid site (H-Z) in a zeolite, desorption of a gaseous and strong base NH_3 can be written as follows:

$$NH_4^+ - Z^- \rightarrow H - Z + NH_3\,(g) \tag{5.8}$$

---

[1] Equation (5.6) is exactly based on the following definitions.

$$BH^+\,(\text{AH sol.}) + A^-\,(\text{AH sol.}) \rightarrow B\,(\text{AH sol.}) + AH\,(l)$$

$$K_a\left(BH^+/AH\right) = \frac{a_{B(\text{AH sol.})}}{a_{BH^+(\text{AH sol.})}\,a_{A^-(\text{AH sol.})}}$$

$$pK_a\left(BH^+/AH\right) = -\log K_a = -\log \frac{a_{B(\text{AH sol.})}}{a_{BH^+(\text{AH sol.})}\,a_{A^-(\text{AH sol.})}}$$

where $x\,(\text{AH sol.})$ shows a solute x in a solvent of AH, and $K_a\left(BH^+/AH\right)$ is the acid dissociation constant of BH^+.

5.1 Paraffin Cracking

On the contrary, a reactant alkane R, a gaseous and weak base, is assumed to form an intermediate RH^+-Z^-, and the reverse step is written as

$$RH^+ - Z^- \rightarrow H - Z + R \text{ (g)} \tag{5.9}$$

Here a relationship equivalent to (5.7) is assumed between the reactions (5.8) and (5.9) on a zeolite. This relationship can be written as (5.10). For any zeolite H-Z and a reactant,

$$\frac{K_a\left(NH_4^+/H-Z\right)}{K_a\left(RH^+/H-Z\right)} = \text{const.} \tag{5.10}$$

As shown in Chap. 2, the standard entropy of desorption of ammonia is approximately constant on various zeolites [16] as well as solid acids such as WO_3/ZrO_2 [17] and SO_4^{2-}/ZrO_2 [18]. Likewise, it is assumed that the desorption entropy of the reactant R is constant, or at least similar, also on various zeolites like that of ammonia. This provides us a linear relationship (5.11) between the desorption enthalpies of ammonia and reactant R on various zeolites.[2]

$$\Delta H\left(RH^+/Z^-\right) = \Delta H\left(NH_4^+/Z^-\right) + \text{const,} \tag{5.11}$$

---

[2] Logarithm of (5.10) gives the following relationships.

$$\ln \frac{K_a\left(NH_4^+/H-Z\right)}{K_a\left(RH^+/H-Z\right)}$$
$$= -\frac{\Delta H\left(NH_4^+/Z^-\right)}{RT} + \frac{\Delta S\left(NH_4^+/Z^-\right)}{R} + \frac{\Delta H\left(RH^+/Z^-\right)}{RT} - \frac{\Delta S\left(RH^+/Z^-\right)}{R}$$
$$= \text{const.}$$
$$\Delta H\left(NH_4^+/Z^-\right) - \Delta H\left(RH^+/Z^-\right)$$
$$= T\left\{\Delta S\left(NH_4^+/Z^-\right) - \Delta S\left(RH^+/Z^-\right)\right\}, +\text{const.}$$

where $\Delta H\left(XH^+/Z^-\right)$ and $\Delta S\left(XH^+/Z^-\right)$ denote the standard enthalpy and entropy, respectively, of the desorption of X from XH^+-Z^-. Here it is assumed that the desorption entropy of the reactant R is constant, or at least similar. This is expressed as follows:

$$\Delta S\left(NH_4^+/Z^-\right) - \Delta S\left(RH^+/Z^-\right) = \beta$$

where β is a constant.
As a result, the equation can be simplified.

$$\Delta H\left(NH_4^+/Z^-\right) - \Delta H\left(RH^+/Z^-\right) = \beta T + \text{const.}$$

This provides a linear relationship (5.11) between the desorption enthalpies of ammonia and reactant R at a fixed temperature on various zeolites. Because the enthalpy change is hardly affected by the temperature (at least in a limited temperature range), availability of eq. (5.11) should not be limited at a fixed temperature.

where $\Delta H\left(\mathrm{NH_4^+/Z^-}\right)$ and $\Delta H\left(\mathrm{RH^+/Z^-}\right)$ denote the standard enthalpies of reactions (5.8) and (5.9), respectively. The former is the ammonia desorption heat, and the latter is the heat of decomposition of $\mathrm{RH^+}$; a constant in (5.11) depends on the reactant alkane.

In the monomolecular mechanism, the formation of a carbonium cation from a Brønsted acid site and an alkane molecule due to (5.1) is the rate-determining step, and therefore, the activation energy experimentally observed (E_a) should be equal to the energy shown by $-\Delta H\left(\mathrm{RH^+/Z^-}\right)$.[3]

$$E_a = -\Delta H(\mathrm{RH^+/Z^-}). \tag{5.12}$$

From (5.11) and (5.12), (5.13) is derived.

$$E_a = -\Delta H\left(\mathrm{NH_4^+/Z^-}\right) + \mathrm{const.} \tag{5.13}$$

A linear relationship between the activation energy of alkane cracking and the ammonia desorption heat with a slope of -1 is shown by this equation. The slope of observed linear relationship in Fig. 5.4 is ca. -1.3, close to -1, indicating the applicability of these assumptions.

The dependence of catalytic activity on the Brønsted acid strength is thus theoretically described based on the measurements with the ammonia IRMS-TPD method. The found principles can be summarized as follows:

- The activation energy of (linear) alkane cracking under the monomolecular conditions is controlled by the Brønsted acid strength; the stronger the acidity, the lower the activation energy, generally resulting in the high activity. The activation energy has a linear relationship with the ammonia desorption heat on the Brønsted acid site. On exceptionally strong acid site (the ammonia desorption heat $> 140\,\mathrm{kJ\,mol^{-1}}$), the relationship is unclear.
- The TOF is also determined by the acid strength, and therefore, the total activity is controlled by the number and strength of Brønsted acid sites.
- This principle is well explained by the Haag–Dessau model of alkane cracking.

5.1.4 Behavior of Acid Sites in 8- and 12-Rings of Mordenite

As shown in Chap. 2, H-mordenite possesses two types of acid sites. The Brønsted acid site in 12-ring is slightly weaker than that in 8-ring. The former is contained in the NaH-mordenite over a wide range of ion exchange degree, while the latter is

---

[3] As $\Delta H\left(\mathrm{RH^+/Z^-}\right)$ is the standard enthalpy of the reaction (5.9), the income of enthalpy by the reaction (5.1) [= the reverse reaction of (5.9) on a zeolite] should be $-\Delta H\left(\mathrm{RH^+/Z^-}\right)$. If the reaction (5.1) is the rate-determining step, the activation energy should be the energy income of this reaction, and hence equal to $-\Delta H\left(\mathrm{RH^+/Z^-}\right)$.

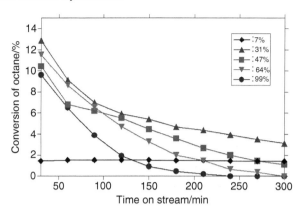

Fig. 5.7 Time course of catalytic activity for octane cracking on in situ NaH-mordenites. The exchange degree ([H]/[Al] = 1−[Na]/[Al]) is shown in *parentheses*. Reproduced with permission from [19]. Copyright American Chemical Society 2005

present only on the sample with a low Na content (high ion exchange degree), as shown in Chap. 3. Catalytic behaviors of these acid sites are significantly different. As shown in Fig. 5.7, a stable activity is observed at 7% of the ion exchange degree ([H]/[Al]), while rapid deactivation is observed at higher ion exchange degrees [19]. It is speculated that the strong acid site in 8-ring forms a coke precursor, and it disperses in both of the 8- and 12-rings to block all the acid sites.

5.2 Adsorption of Aromatic Hydrocarbons

Adsorption of such an aromatic compound as toluene on zeolites gathers attention from multiple viewpoints as follows, and TPD of the aromatic hydrocarbons has been used to study their adsorption properties [20].

Ammonia can be used as a probe molecule to analyze the acidic property of solid, as reviewed in this book. The adsorption of ammonia, as well as adsorption of other molecules, can be affected by steric hindrance or, in contrast, "confinement effect" [13] in a small cavity. These steric effects disturb the estimation of intrinsic acid strength from the adsorption heat of a probe. From this sense, it is proposed that a small alkaline cation, e.g., Na^+, should be an ideal base probe for the acidity evaluation of a solid surface to avoid from any steric effects. An aromatic hydrocarbon such as toluene is strongly adsorbed on Na^+ in zeolites [21], and it may be possible to study the nature of interaction between the acid site (or ion-exchange site) and Na^+ through the adsorption behavior.

The adsorption of aromatic compound is also a subject of research on diffusion of molecules in micropore [22], because the molecular size is similar to the micropore size. Moreover, the adsorption of aromatic compound can be a tool for probing the cation position in very narrow pores [23].

Apart from the above applications as a probe to investigate the physico-chemical properties of zeolite, the adsorption of such a hydrocarbon compound as toluene, itself, gathers attention. Nowadays, efforts are being made to control the emission of unburned hydrocarbon and NO_x from automobile engines during "cold-start period" when catalyst temperature has not reached to a suitably high level. It is promising for the removal of the hydrocarbons to store them at a low temperature and release to send them to the combustion catalyst after it is heated [24, 25]. In addition, for the removal of NO_x, the storage of hydrocarbons as a reductant is useful [26]. Moreover, toluene is a typical volatile organic compound (VOC), and removal of it from atmosphere by catalytic combustion or adsorption is being attempted widely [27].

In this section, the quantitative analysis of toluene TPD on Na-zeolites [28] is reviewed.

Figure 5.8 shows the change in TPD spectrum with varying the W/F ratio (see Chap. 2) on a sample of Na-ZSM-5. The TPD of toluene on a Na-zeolite shows l-(low temperature) and h- (high temperature) peaks. At a low W/F, the l-peak is not observed, because the weakly adsorbed toluene is purged during the preevacuation before the TPD measurement. The peak temperature of h-peak shifts with varying the W/F ratio, as observed in the ammonia TPD on an H-zeolite. A linear relationship between $\ln T_m - \ln \frac{A_0 W}{F}$ and $\frac{1}{T_m}$ is observed in Fig. 5.9. This indicates that the

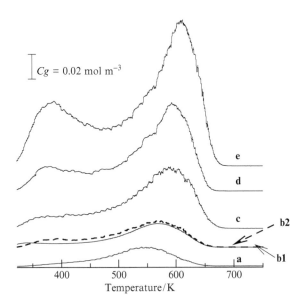

Fig. 5.8 Toluene TPD profiles on Na-ZSM-5 ($SiO_2/Al_2O_3 = 24$) at different W/F ratios; (**a**) $W = 0.1$ g, $F = 400 \text{ cm}^3 \text{ min}^{-1}$; (**b1**) and (**b2**) $W = 0.1$ g, $F = 200 \text{ cm}^3 \text{ min}^{-1}$; (**c**) $W = 0.1$ g, $F = 100 \text{ cm}^3 \text{ min}^{-1}$; (**d**) $W = 0.3$ g, $F = 200 \text{ cm}^3 \text{ min}^{-1}$; (**e**) $W = 0.5$ g, $F = 200 \text{ cm}^3 \text{ min}^{-1}$. The experiments at $W = 0.1$ g, $F = 200 \text{ cm}^3 \text{ min}^{-1}$ were repeated as (**b1**) and (**b2**) under the same conditions. The measurements were carried out under atmospheric conditions

5.2 Adsorption of Aromatic Hydrocarbons

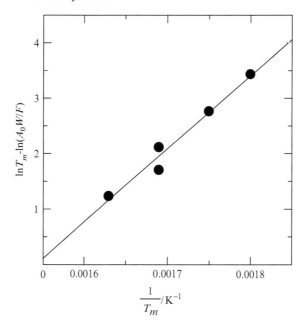

Fig. 5.9 Relationship between $\ln T_m - \ln(A_0 W/F)$ and $1/T_m$ on Na-ZSM-5 ($SiO_2/Al_2O_3 = 24$)

TPD process under these conditions is controlled by the equilibrium, and the readsorption of toluene from the gas phase freely occurs, as observed in the ammonia TPD on solid acids.

Moreover, ΔS (standard entropy change of desorption) is calculated to be 148 J K^{-1} mol^{-1} from the slope and intercept of the linear relationship in Fig. 5.9. The entropy consists of entropy changes due to the phase transformation [ΔS_{trans}] and the mixing [$\Delta S_{mix}(T)$]. The latter term is calculated to be about 60 J K^{-1} mol^{-1} from the gaseous composition at the peak maximum, and the former is hence estimated to be ca. 90 J K^{-1} mol^{-1}. Trouton's rule states that the entropy change with respect to the liquid vaporization is approximately constant (in many cases 80–110 J K^{-1} mol^{-1}, and 87.3 J K^{-1} mol^{-1} for toluene) for various materials, showing that the entropy change is mainly determined by the free volume of a gas molecule [29]. The entropy change for desorption can also be determined by the free volume of gas molecule. Therefore, the entropy change due to the phase transformation estimated by the present study (ca. 90 J K^{-1} mol^{-1}), which agrees with the vaporization entropy change shown by Trouton's rule, supports the validity of the present analysis.

A similar value of entropy has been obtained in the ammonia TPD on various solid acids, as shown in Chap. 2. This indicates

- The same analytical method can be applied to both the ammonia TPD on the solid acid and the toluene TPD on the Na-zeolite.
- Validity of the both experiments is well supported by thermodynamics.

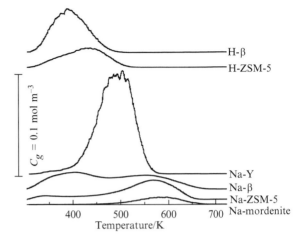

Fig. 5.10 TPD profiles of toluene over various zeolites; sample amount 0.1 g; He flow rate 200 cm³ min⁻¹; heating rate 5 K min⁻¹

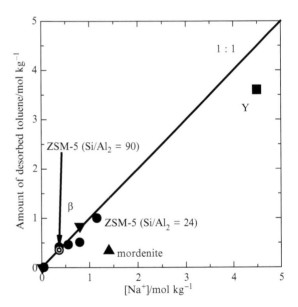

Fig. 5.11 Relationship between the amounts of desorbed toluene and Na⁺ on ZSM-5 (*filled circle* at Si/Al$_2$ = 90 and *open double circle* at Si/Al$_2$ = 24), mordenite (*open triangle*, Si/Al$_2$ = 15), Y (*open square*, Si/Al$_2$ = 5) and β (*open inverted triangle*, Si/Al$_2$ = 25)

From the toluene TPD spectra on various Na-zeolites shown in Fig. 5.10, the number of toluene molecules showing the *h*-peak is calculated. Figure 5.11 shows that the number of toluene molecules equals the number of Na atoms on most of the Na-zeolites. It demonstrates that toluene is strongly adsorbed on the Na⁺ cation in the Na-zeolite. As an exception, a Na-MOR shows the number of toluene molecules smaller than the number of Na atoms, presumably because the adsorption on the Na⁺ cation located in the 8-ring is suppressed or weakened by steric hindrance.

The heat of toluene desorption is calculated by the curve-fitting method derived in Chap. 2. The heat was in the order of MOR > MFI > BEA > FAU. This is in agreement with the order of heat of ammonia desorption on the corresponding

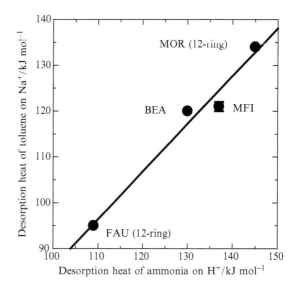

Fig. 5.12 Relationship between heats of toluene desorption on Na$^+$ and ammonia desorption on H$^+$ in corresponding zeolite

H-zeolite, as shown in Fig. 5.12. As shown in Chap. 4, the electron withdrawing nature of ion exchange site is determined by the framework structure of zeolite. This electron-withdrawing property is supposed to control the two chemical natures. One is the acid strength, namely the positive charge of proton in the H-form zeolite to give the strong interaction between a base and the H-zeolite. Another is a more positive charge of Na$^+$ cation in the Na-form, providing the strong interaction between toluene and the Na-zeolite. The interaction presumably exists between the π-electrons of aromatic ring and the empty orbital of Na$^+$ cation, and therefore, it is believed that the similar adsorption behaviors are observed also for other aromatic compounds.

In summary, the following knowledge is obtained by these studies.

- An aromatic hydrocarbon molecule is strongly adsorbed by a Na$^+$ cation in a Na-zeolite.
- The adsorption strength is controlled by the electron-withdrawing nature of zeolite framework, as well as the acid strength of the corresponding H-form zeolite.

5.3 Friedel–Crafts Alkylation on Ga-MCM-41

Mesoporous[4] silicas such as FSM-16 [30] and MCM-41 [31] have been synthesized in the early of 1990s. It was expected that introduction of Al into such materials yielded a highly active catalyst for cracking of large molecule, because the

---

[4] Definition of "mesopore" is pores with diameters 2–50 nm. The smaller pore is called micropore, while the larger one is called macropore.

accessibility to the active site on the mesopore wall should be highly compared to that in the conventional zeolite, namely, microporous material. However, based on the knowledge about the generation of Brønsted acidity studied through the previous chapters, it has to say that the strong Brønsted acidity is generated only in specific framework topologies. Although many efforts have been paid, it has been difficult to generate strong Brønsted acidity on the mesopore wall of such a material with amorphous structure. However, other applications have been found. Here an example is shown.

Introduction of Ga with a gel equilibrium adjustment method [32] gives a gallium-containing MCM-41 (referred as Ga-MCM-41 hereafter). Similarly to an MFI type gallosilicate, all Ga atoms are in the tetra-coordinated environment, as shown by EXAFS (extended X-ray adsorption fine structure) spectrum of Ga K-edge (Fig. 5.13c). However, after the conversion into H-form via ion-exchange with an NH_4 salt and calcination, the coordination number is reduced (Fig. 5.13d), indicating that a fraction of Ga moves onto the external surface (mesopore wall). Infrared spectrum of adsorbed pyridine shows that most of the acid site generated by such an extra-framework Ga species is Lewis type. Impregnation of Ga gives the similar Ga species mainly on the external surface and Lewis acidity, as shown in Fig. 5.13e. Ammonia MS-TPD shows the presence of very strong acid sites (large desorption peaks at high temperatures) on both Ga-MCM-41 and Ga-impregnated MCM-41 (Fig. 5.14a, b). At a low Ga content, the number of acid site was close to the number of Ga atom, whereas the acid amount looks saturated in the region [Ga] > 0.4 mol kg^{-1}, as shown in Fig. 5.15.

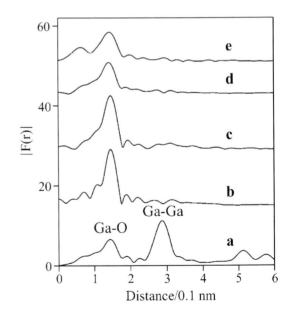

Fig. 5.13 Fourier transform of k^3-weighted Ga K-edge EXAFS spectra; (**a**) β-Ga$_2$O$_3$, (**b**) MFI gallosilicate (H-Form, [Ga] = 0.24 mol kg^{-1}), (**c**) Ga-MCM-41 (as-synthesized form, [Ga] = 0.55 mol kg^{-1}), (**d**) Ga-MCM-41 (H-form, [Ga] = 0.55 mol kg^{-1}), and (**e**) Ga-impregnated MCM-41 ([Ga] = 0.45 mol kg^{-1}). Reproduced with permission from [33]. Copyright Elsevier 2001

5.3 Friedel–Crafts Alkylation on Ga-MCM-41

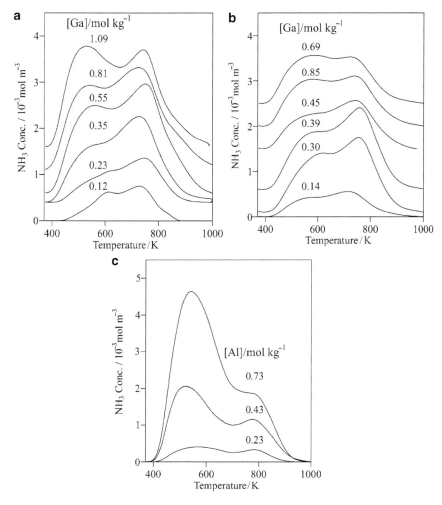

Fig. 5.14 Ammonia MS-TPD spectra of (**a**) Ga-MCM-41, (**b**) Ga-impregnated MCM-41, and (**c**) Al-MCM-41. Reproduced with permission from [33]. Copyright Elsevier 2001

On the Al-introduced MCM-41, also Lewis acidity is observed, and its strength should be weak, as apparently from the MS-TPD peak at a low temperature (Fig. 5.14c).

Friedel–Crafts alkylation of benzene with benzyl chloride (5.14) in a liquid phase proceeds on the Lewis acid site, as shown in Fig. 5.16.

$$\text{C}_6\text{H}_5\text{-CH}_2\text{Cl} + \text{C}_6\text{H}_6 \longrightarrow \text{C}_6\text{H}_5\text{-CH}_2\text{-C}_6\text{H}_5 + \text{HCl} \quad (5.14)$$

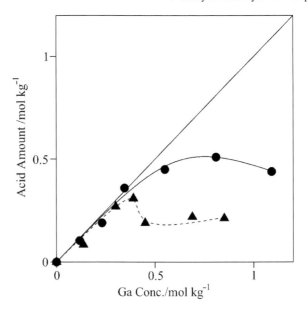

Fig. 5.15 Acid amount determined by ammonia MS-TPD on Ga-MCM-41 (*filled circle*) and Ga-impregnated MCM-41 (*filled triangle*). Reproduced with permission from [33]. Copyright Elsevier 2001

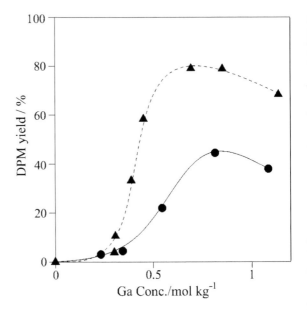

Fig. 5.16 Catalytic activity for Friedel–Crafts alkylation of benzene with benzylchloride into diphenylmethane (DPM) on Ga-MCM-41 (*filled circle*) and Ga-impregnated MCM-41 (*filled triangle*). Reproduced with permission from [33]. Copyright Elsevier 2001

However, it is not simple to interpret the change in activity with varying the preparation method and Ga content. The activity is higher on Ga-impregnated MCM-41 than Ga-MCM-41. The acid amount is higher on Ga-MCM-41, which is inconsistent with the catalytic activity. The activities on the both series of catalysts show a marked increase at ca. $0.4\,\mathrm{mol\,kg^{-1}}$ of the Ga content, where the increase of acid amount is saturated. Not only the strong Lewis acidity but also additional character of catalyst seems to be required for this kind of reaction. It is speculated that agglomeration of Ga is necessary to accelerate the reaction [33].

Further study on the relationship between the acidic property and activity resulted in findings of new heteropoly acid-related solid acid catalysts such as $H_4NbW_{11}O_{40}/WO_3$-Nb_2O_5 [34] and $H_9P_2W_{15}Nb_3O_{62}/SiO_2$ [35] active for various liquid-phase reactions.

5.4 Amination of Phenol into Aniline on Ga/ZSM-5

Aniline is an industrially important chemical substance and has been produced by amination of phenol (5.15).

$$C_6H_5\text{-OH} + NH_3 \rightarrow C_6H_5\text{-}NH_2 + H_2O \qquad (5.15)$$

This reaction has been carried out in the gas phase under a high pressure of ammonia with alumina catalyst [36, 37]. Researches to find a more active catalyst have been continued in order to suppress the energy consumption for operation under the high pressure and high temperature conditions. H-form zeolites such as H-ZSM-5 [38] and H-β [39] have been found to possess relatively high activities, and therefore, Brønsted acidity should be necessary for this reaction.

Katada et al. found a high activity of Ga-containing silicate for this reaction. Compared to zeolites such as H-ZSM-5 and H-β, MFI type gallosilicate shows obviously higher activity, as shown in Fig. 5.17. Dependence of the activity on the Ga content is unclear, but the activity is observed only on the samples with a high Ga content ([Ga] $> 0.5\,\mathrm{mol\,kg^{-1}}$) [40].

Structural analysis based on the d-spacing of (10 0 0) plane, micropore volume, and crystal morphology shows that Ga is easily incorporated into the framework in the low Ga concentration region up to ca. $0.4\,\mathrm{mol\,kg^{-1}}$, but further Ga incorporation is difficult. Ammonia MS-TPD (Fig. 5.18) shows a desorption peak around 600 K with a shoulder at ca. 800 K. The former is attributed to the acid site generated by isomorphous substitution of Si by Ga, because the number of this type of acid site is almost equal to the number of Ga atom at [Ga] $< 0.4\,\mathrm{mol\,kg^{-1}}$, as shown in Fig. 5.19. On the contrary, the latter is ascribed to the extra-framework Ga species, because the latter increased significantly at [Ga] $> 0.4\,\mathrm{mol\,kg^{-1}}$ [41].

From these findings it is concluded that the activity for amination of aniline on the gallosilicate is ascribed to the combination of extra-framework Ga species and

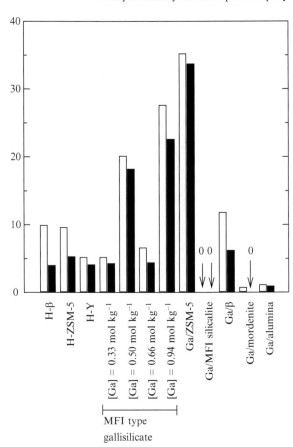

Fig. 5.17 Comparison of catalytic activity for amination of phenol after 0.5–1 (*white*) and 3.5–4 h (*black*) of the time on stream

Brønsted acid site, but not to the Brønsted acid site generated by the isomorphous substitution only. The strength of Brønsted acid site due to the Ga in MFI framework ($\Delta H = 130\,\text{kJ mol}^{-1}$) is similar to that due to the Al in MFI framework [41], supporting that the exceptionally high activity of gallosilicate is not simply due to the Brønsted acid site only.

A more active catalyst can be designed based on this information. The above findings indicate that the combination of Ga species and Brønsted acid site with $130\,\text{kJ mol}^{-1}$ of the ammonia desorption heat generated a high activity. The latter can be more easily obtained by Al-containing MFI silicate, i.e., ZSM-5 zeolite. It is expected that the impregnation of Ga on ZSM-5 yields an active catalyst. Figure 5.17 shows a remarkable activity of Ga/ZSM-5 for the amination of phenol [40]. This catalyst shows an extremely long life, as shown in Fig. 5.20. The role of extra-framework Ga species is the acceleration of NH_2 formation, as discussed from kinetic and spectroscopic analyses [42]. Useful information from the ammonia TPD thus makes us to design a new catalyst.

5.4 Amination of Phenol into Aniline on Ga/ZSM-5

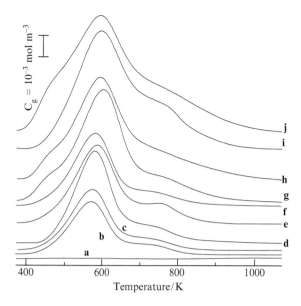

Fig. 5.18 Ammonia MS-TPD spectra of silicalite (**a**) and H-gallosilicate [the Ga content (mol kg^{-1}), (**b**): 0.13, (**c**): 0.17, (**d**): 0.21, (**e**): 0.25, (**f**): 0.26, (**g**): 0.31, (**h**): 0.48, (**i**): 0.60, and (**j**): 0.81] after water vapor treatment

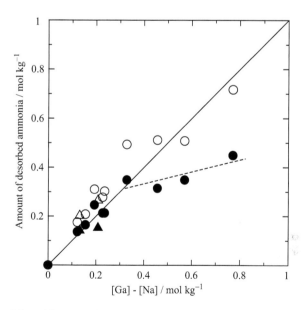

Fig. 5.19 Plot of the acid amount determined from the whole peak area (*open circle* and *open triangle*) and from the simulated *h*-peak (*filled circle* and *filled triangle*) against [Ga]−[Na] content over H- (*open circle* and *filled circle*) and HNa-type (*open triangle* and *filled triangle*) gallosilicate

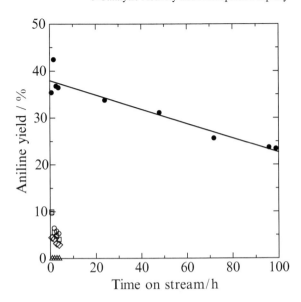

Fig. 5.20 Catalytic activity and its time course on alumina (*open triangle* JRC-ALO4, *open diamond* Neobeads), H-β (*open square*), H-ZSM-5 (*open circle*) and Ga/ZSM-5 (*filled circle*, [Ga] = 0.3 mol kg$_{\text{-zeolite}}^{-1}$)

References

1. W.O. Haag, R.M. Dessau, in *Proceedings of the 8th International Congress on Catalysis*, vol. 2, (Varlag-Chemie, Weinheim, 1984), pp. 305–316
2. S.J. Collins, P.J. O'Malley, J. Catal. **153**, 94 (1995)
3. C. Seitz, A.L.L. East, J. Phys. Chem. A **106**, 11653 (2002)
4. M. Boronat, A. Corma, Appl. Catal. A Gen. **336**, 2 (2008)
5. T. Hashiba, D. Hayashi, N. Katada, M. Niwa, Catal. Today **97**, 35 (2004)
6. N. Katada, K. Suzuki, T. Noda, W. Miyatani, F. Taniguchi, M. Niwa, Appl. Catal. A Gen. **373**, 208 (2010)
7. B.A. Williams, S.M. Babitz, J.T. Miller, R.Q. Snurr, H.H. Kung, Appl. Catal. A Gen. **177**, 161 (1999)
8. M. Kuehne, H.H. Kung, J.T. Miller, J. Catal. **171**, 293 (1997)
9. J.A. Van Bokhoven, M. Tromp, D.C. Koningsberger, J.T. Miller, J.A.Z. Pieterse, J.A. Lercher, B.A. Williams, H.H. Kung, J. Catal. **202**, 129 (2001)
10. N. Katada, Y. Kageyama, K. Takahara, T. Kanai, H.A. Begum, M. Niwa, J. Mol. Catal. A Chem. **211**, 119 (2004)
11. S. Kotrel, M.P. Rosynek, J.H. Lunsford, J. Phys. Chem. B **103**, 818 (1999)
12. F. Eder, J.A. Lercher, J. Phys. Chem. B **101**, 1273 (1997)
13. G. Sastre, A. Corma, J. Mol. Catal. A Chem. **305**, 3 (2009)
14. N.F. Hall, J. Am. Chem. Soc. **52**, 5115 (1930)
15. L.P. Hammet, A.J. Deyrup, J. Am. Chem. Soc. **54**, 2721 (1932)
16. M. Niwa, N. Katada, M. Sawa, Y. Murakami, J. Phys. Chem. **99**, 8812 (1995)
17. N. Naito, N. Katada, M. Niwa, J. Phys. Chem. B **103**, 7206 (1999)
18. N. Katada, J. Endo, K. Notsu, N. Yasunobu, N. Naito, M. Niwa, J. Phys. Chem. B **104**, 10321 (2000)
19. M. Niwa, K. Suzuki, N. Katada, T. Kanougi, T. Atoguchi, J. Phys. Chem. B **109**, 18749 (2005)
20. N. Sivasankar, S. Vasudevan, J. Phys. Chem. B **108**, 11585 (2004)
21. V.R. Choudhary, K.R. Srinivasan A.P. Singh Zeolites **10**, 16 (1990)
22. T. Masuda, Y. Fujitaka, T. Nishida, K. Hashimoto, Microporous Mesoporous Mater. **23**, 157 (1998)

References

23. M. Kato, K. Itabashi, A. Matsumoto, K. Tsutsumi, J. Phys. Chem. B **107**, 1788 (2003)
24. K.F. Czaplewski, T.L. Reitz, Y.J. Kim, R.Q. Snurr, Microporous Mesoporous Mater. **56**, 55 (2002)
25. D.S. Lafyatis, G.P. Ansell, S.C. Bennet, J.C. Frost, P.J. Millington, R.R. Rajaram, A.P. Walker, T.H. Ballinger, Appl. Catal. B Environ. **18**, 123 (1998)
26. R. Yoshimoto, T. Ninomiya, K. Okumura, M. Niwa, Appl. Catal. B Environ. **75**, 175 (2007)
27. S.-W. Baek, J.-R. Kim, S.-K. Ihm, Catal. Today **93**, 575 (2004)
28. R. Yoshimoto, K, Hara, K. Okumura, N. Katada, M. Niwa, J. Phys. Chem. C **111**, 1474 (2007)
29. G.M. Barrow, *Physical Chemistry*, 5th edn. (McGraw Hill New York, 1988)
30. T. Yanagisawa, T. Shimizu, K. Kuroda, C. Kato, Bull. Chem. Soc. Jpn. **63**, 988 (1990)
31. C.T. Kresge, M.E. Leonowicz, W.J. Roth, J.C. Vartuli, J.S. Beck, Nature **359**, 710 (1992)
32. R. Ryoo, J.M. Kim, J. Chem. Soc. Chem. Commun. **711** (1995)
33. K. Okumura, K. Nishigaki, N. Niwa, Microporous Mesoporous Mater. **44–45**, 509 (2001)
34. K. Okumura, K. Yamashita, K. Yamada, M. Niwa, J. Catal. **245**, 75 (2007)
35. K. Okumura, S. Ito, M. Yonekawa, A. Nakashima, M. Niwa, Top. Catal. **52**, 649 (2009)
36. M. Yasuhara, F. Matsunaga, Eur. Patent 321275 A2, 1989
37. A.A. Schutz, L.A. Cullo, PCT Patent WO9305010 A1, 1993
38. C.D. Chang, W.H. Lang, Eur. Patent 62542 A1, 1982
39. N. Katada, S. Iijima, H. Igi, M. Niwa, Stud. Surf. Sci. Catal. **105**, 1227 (1997)
40. N. Katada, S. Kuroda, M. Niwa, Appl. Catal. A Gen. **180**, L1 (1999)
41. T. Miyamoto, N. Katada, J.-H. Kim, M. Niwa, J. Phys. Chem. B **102**, 6738 (1998)
42. N. Katada, T. Doi, T. Shinmura, S. Kuroda, N. Niwa, Stud. Surf. Sci. Catal. **145**, 197 (2003)

Chapter 6
CVD of Silica for the Shape Selective Reaction

Abstract Chemical vapor deposition (CVD) of silicon alkoxide on the external surface of zeolite is proposed to realize the shape selective reaction and adsorption. Formation, characterization and control function of the silica overlayer on zeolites, mordenite, ZSM-5, A, and Y-zeolites are elucidated. The fine control of the pore-opening size of zeolite is an example of the molecular engineering of zeolites.

6.1 Reactants and Products Shape Selectivity, Concept and Definition

Shape selectivity is a concept for the selective formation of the desired product based on the zeolite catalysts. The generation of the shape selectivity is realized only with the micro porosity of zeolites effectively utilized. Various energy-saving processes of the catalytic reaction and the adsorption separation are realized by utilizing the excellent property of the shape selectivity on zeolites. Thus, the enhancement of the shape selectivity has been studied extensively from various view-points. The study on the selective production using the modified zeolites is a subject which is most typical on the zeolite catalysts. It includes utilizations of various available reagents, effective modification processing, and preparation mechanism.

Shape selectivity is divided into following three categories, as shown in Fig. 6.1. Reactant selectivity is defined as that any molecule which cannot enter into the pore does not react. In a similar way, product selectivity is defined as that any molecule which cannot be desorbed from the pore is not produced. Transition-state selectivity is not usual but has been indicated in a following, i.e., any product of which intermediate in the transition state is not formed inside the pore is not formed [1]. The latter selectivity should be caused by the three dimensional inherent structure of the zeolite, and will not be a subject in the present description. Improvement of the reactant and product selectivity based upon the control of pore-opening size is reviewed in this chapter.

6.2 Chemical Vapor Deposition of Silica and the Procedure

Niwa et al. have found that chemical vapor deposition of silica on the external surface of zeolite remarkably (in some cases, perfectly) enhances the shape selectivity [2]. The chemical vapor deposition (CVD) method has been studied extensively on various zeolites. Usually, silicon tetra-alkoxide, i.e., $Si(OCH_3)_4$ or $Si(OC_2H_5)_4$,

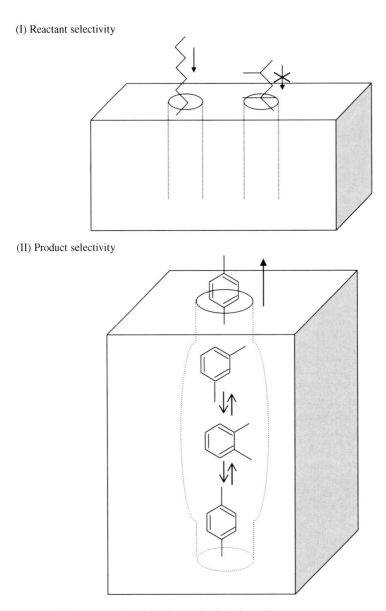

Fig. 6.1 Three categories of the shape selectivity in zeolites

(III) Transition state selectivity

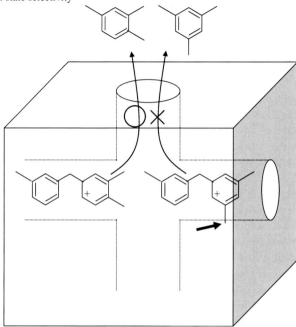

Fig. 6.1 (continued)

is utilized as a reagent to modify the zeolite property. These molecules are characterized as the reagent for this method, as mentioned in a following:

- The alkoxide molecule cannot enter the pore of zeolite, because the molecular size is larger than the pore size of zeolite. Thus, the alkoxide is deposited only on the external surface. Inside of the zeolite pore is retained, but the external surface is modified. However, this principle is not applied to Y-zeolite exceptionally, as mentioned in Sect. 6.4.4.
- The deposition of the molecule on the external surface occurs readily, because it reacts with the hydroxide on the zeolite rapidly.
- The elongation of the Si–O bond takes place through the formation of siloxane bond, Si–O–Si, because four reacting alkoxide groups are attached. Silica network is formed due to the multiple functionality of the alkoxide.
- The formed silica is stabilized on the zeolite surface, and therefore the high stability of the modified material is anticipated. Similarity of the structures in zeolite and silica may be a fundamental reason.
- $Si(OCH_3)_4$ is deposited on zeolites more readily than $Si(OC_2H_5)_4$. No other preferential property has been indicated for both molecules.

Silane SiH_4 was used for modifying the zeolite property by Vansant et al. before utilization of these alkoxides [3]. SiH_4 modifies the zeolite totally, not only on the external but in the internal of zeolite because of the small molecular size.

Si_2H_6 as well as SiH_4 modifies the zeolite to change the adsorption property. The nonselective modification is discriminated clearly from the CVD by silicon alkoxide molecules. Handling of these silane compounds is however expected to be difficult, because these are explosive chemical compounds.

Alkoxide derivatives are used to create the property of modified zeolites which is different from those modified with $Si(OR)_4$. $Si(OCH_3)_x(CH_3)_{4-x}$, where $x = 3, 2$, and 1, are studied for the CVD on HZSM-5 [4].

In place of the silicon compounds, $Ge(OCH_3)_4$ is available for the same purpose [5,6]. Germanium oxide deposited on the zeolite, however, easily changes the structure under the humidified conditions to form the agglomerated large particle of germanium oxide. Thus, no preferential property for this compound is confirmed in compared with $Si(OR)_4$. Therefore, the germanium compound is used only for the spectroscopic study on the deposited oxide to discriminate between the deposited oxide (Ge) and the zeolite (Si).

An apparatus used in the laboratory for the CVD is shown in Fig. 6.2. It is a glass-made apparatus installed with the vacuum equipments to measure the amount of deposited silica precisely by means of a quartz microbalance. This method is recommended for a quantitative study on the CVD of silica, because the amount of silica deposited is measured in situ precisely. It is difficult to quantify the deposited Si precisely after the deposition, because the amount of the silicon compound deposited is so small compared to the zeolite bulk. This method is however not recommended for the production of a large amount of the modified zeolite. Studies on the production of the CVD zeolites for the practical utilization are elucidated in Sect. 7.1.3 in Chap. 7.

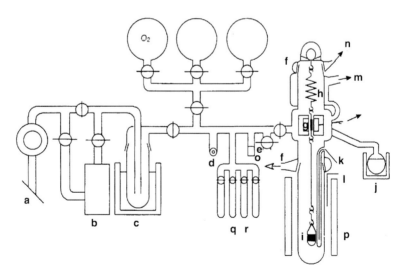

Fig. 6.2 Apparatus for CVD in the laboratory: (**a**) rotary vacuum pump; (**b**) diffusion vacuum pump; (**c**) liquid nitrogen trap; (**d**) pressure transducer; (**e**) Pirani gauge; (**f**) cooled connector; (**g**) displacement meter; (**h**) quartz spring balance; (**i**) sample; (**j**) boiled CCl_4; (**k, l**) thermocouple; (**m, n**) to Graham condenser; (**o**) circulated coolant: (**p**) electric furnace; (**q, r**) $Si(OCH_3)_4$ and adsorbate liquids

6.2 Chemical Vapor Deposition of Silica and the Procedure

Fig. 6.3 Preparation of silica deposited H-mordenite using the apparatus in Fig. 6.2; *evac* evacuation; *ad* admission of the alkoxide

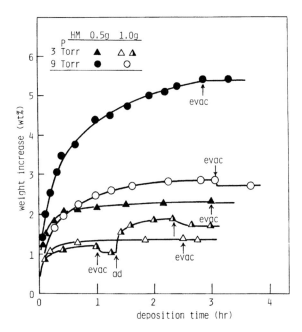

A typical procedure of the preparation of CVD zeolite using the apparatus by varying the amount of zeolite and the partial pressure of alkoxide is shown in Fig. 6.3. Zeolite sample is first evacuated at such a high temperature as 673 K to remove sorbed water. Gaseous silicon alkoxide is allowed to be deposited on the evacuated sample from the liquid reservoir kept at such a constant temperature as 273 K. The weight of the sample increases upon deposition of silica, and the resulting expansion length of quartz micro balance is monitored by a displacement meter. The change of micro balance expansion is received by the meter, and the measured amplitude in DC voltage is converted into the digital signal through an AD (analogue-digital) converter to transfer into a personal computer. After proceeding of the CVD for a while, the deposition seems to be discontinued; however, this is not because the zeolite surface is inactivated, but because the alkoxide molecule is not transported to the zeolite sample due to the block of the alkoxide diffusion by the produced molecules, which is so-called "Vapor lock." As shown in Fig. 6.3, the weight increase in wt % is controlled by the weight of zeolite and the partial pressure of alkoxide. The weight increase is discontinued when almost all the alkoxide in the glass vessel is deposited on the zeolite, because the amount of silica deposited is roughly equal to that of the alkoxide molecules in the glass vessel. Therefore, the alkoxide is readmitted after pumping the gaseous products to continue the deposition. Depositions of silica are repeated to arrive at the required weight increase of silica. Finally, the deposited carbon containing residue is removed by the calcination with oxygen. The amount of deposited silica is measured precisely from the weight gain in the final step.

6.3 Formation of Silica Overlayer on the External Surface

6.3.1 Method of Benzene-Filled Pore for the Measurement of External Surface Area

Measurement of the external surface area on a zeolite is required to evaluate the thickness of silica layer deposited quantitatively. The averaged thickness of silica layer is measured from the external surface area and the T-site (tetrahedral site occupied by Si or Al) density. The quantitative evaluation of the deposited silica layer is important in this study.

A method of benzene-filled pore is recommended to measure the external surface area [7]. Benzene is adsorbed on the dried zeolite at room temperature, until the pore is fully filled with benzene. Usually, it takes about 12 h for the pores to be filled with benzene. Nitrogen is adsorbed on the benzene-filled sample at liquid nitrogen temperature 77 K under the conditions of a partial pressure of nitrogen P/P^0, ca. 0.3, and the amount of nitrogen gas desorbed at the temperature of dry ice in ethanol is measured with a thermal conductivity detector. A continuous flow method with He–N$_2$ mixture (N$_2$ volume content, ca.30%) carrier gas is useful to do the experiment. Because the pore of zeolite is filled with benzene, nitrogen is adsorbed only on the external surface. One-point method of BET surface area measurement conducted in a continuous-flow apparatus is modified to measure the external surface area. Therefore, the amount of nitrogen to be adsorbed in a monolayer (V_m) is related with the desorbed amount (V_0) in a following,

$$V_m = V_0 \left(1 - \frac{P}{P^o}\right). \tag{6.1}$$

The surface area is calculated from number of N$_2$ molecule adsorbed in a monolayer and the cross-sectional surface area (0.162 nm^2).

Water is adsorbed on the hydrophobic zeolite but not sufficiently on the hydrophilic zeolite. On the other hand, such an alkane as hexane is not adsorbed fully on the hydrophobic zeolite. Benzene has the property which is intermediate between water and alkane, and the adsorption amount does not largely depend on the silica to alumina ratio of a zeolite HZSM-5. Benzene molecule is therefore selected as the adsorbent from its hydrophobic and hydrophilic properties, and the method of benzene-filled pore is applicable to zeolites with low-to-high Si/Al ratios. Applicability of this method to various zeolites is confirmed experimentally. External surface area as measured by this method is confirmed by the scanning electron microscopy (SEM) geometrical observation.

6.3.2 Mechanism of CVD to form the Silica Overlayer

$Si(OCH_3)_4$ is first anchored on the zeolite as the silicon-trialkoxide, i.e.,

$$Si(OCH_3)_4 + Z-OH \rightarrow Z-O-Si(OCH_3)_3 + CH_3OH \qquad (6.2)$$

This initiation reaction takes place on the hydroxide of the external surface, since the IR observation shows that the isolated SiOH at $3,745\,cm^{-1}$ diminish upon the deposition, and the isolated silanol is believed to be located on the external surface [8]. The reaction (6.2) seems to occur even at room temperature, but the product methanol remains adsorbed in the zeolite.

Subsequent reactions are not simple to be studied, but seem to depend on the zeolite acidity and the deposition temperature. On the acid type zeolite at the temperature above 573 K, the product methanol is converted into dimethylether and hydrocarbons. Water is by-produced simultaneously, and activates the deposited residue to generate the hydroxide for the subsequent deposition, i.e.,

$$CH_3OH \rightarrow CH_3OCH_3 \text{ and } H_2O \rightarrow C_nH_{2n}, C_nH_{2n+2}, \text{ and } H_2O \qquad (6.3)$$

$$Z-O-Si(OCH_3)_3 + H_2O \rightarrow Z-O-Si(OH)(OCH_3)_2 + ROH \qquad (6.4)$$

Silicon alkoxide is deposited on the produced hydroxide, and the siloxane bond elongates. Condensation reaction from the alkoxide molecules to form the polymerized silicon compound and dimethylether also seems to occur, i.e.,

$$Z-O-Si(OCH_3)_3 + Si(OCH_3)_4 \rightarrow Z-O-Si(OCH_3)_2-OSi(OCH_3)_3 + CH_3OCH_3 \qquad (6.5)$$

Thus, the silica network is formed on the external surface due to the reactions (6.4) and (6.5). The formed silica network covers the zeolite to change the property.

On the other hand, the deposition of $Si(OCH_3)_4$ on Na-zeolite is simple in compared to on the H-zeolite even at such a high temperature as 593 K [9]. $Si(OCH_3)_4$ is deposited accompanied by the formation of methanol, as shown by (6.2). No other products than methanol are formed, because no acid site exists. The weight gain due to the deposition of silica is saturated easily on the Na-type zeolite. The saturated surface density of silicon on the Na-mordenite is ca. $10\,Si\,nm^{-2}$ from the amount of deposited silica and the external surface area. The zeolite surface inactivated by the saturated deposition is, however, activated by the admission of water at 593 K, but the degree of weight increase by further deposition becomes small gradually. The saturation behavior and the amount of silica at the saturation are nearly independent of the temperature, from room temperature to 593 K. The surface concentration, $10\,Si\,nm^{-2}$ on the mordenite, may suggest the formation of silica monolayer, because the T-site density of mordenite on the (001) plane is $8.6\,nm^{-2}$.

Thus, it is important to notice that chemical vapor deposition of $Si(OCH_3)_4$ on Na-zeolite may readily lead to a formation of silica monolayer on the external surface to arrive at the saturation conditions. Also on the H-mordenite at room temperature, the deposition terminates at 14 Si nm^{-2} of the surface concentration, which is a little larger than the site density in the monolayer. Thus, it becomes obvious that the deposition to form the monolayer readily proceeds, but the further accumulation of the layer does not occur rapidly. On the other hand, the formation of multiple layers proceeds on the H-type zeolite at above 573 K, because water is formed simultaneously as a result of dehydration of methanol to activate the deposited residue. Difference between the depositions on Na- and H-zeolites is therefore obvious in the step of multiple-layer formation. Thus, careful considerations are required to prepare the thick silica layer on the external surface of zeolites, because number of the silica layer required for the control depends on the zeolite. Mono to triple layer of silica is enough to control the mordenite pore, but more than 20 layers are needed for the silicalite.

6.3.3 Formation of Silica Overlayer on Zeolite and Metal Oxide, and Its Function

An application of the silica CVD to metal oxides to prepare the molecular sieving silica overlayer is explained, as shown in Fig. 6.4 [10]. Silica monolayer is formed by the deposition of silica on metal oxides such as Al_2O_3, TiO_2, ZrO_2, and SnO_2 that have the weak basicity. Aldehyde (e.g., butanal, benzaldehyde and 1-naphthaldehyde) is adsorbed as carboxylate anion on the surface of these metal oxides of which the surface density is limited to ca. 50%, silica is deposited on the remaining bared surface, and thus the surface is totally covered by the carboxylate anion and the silica. Subsequently, adsorbed carboxylate anion is removed by the reaction with ammonia into nitrile to create the vacant (pore) site of silica of which size is adjusted by the shape of the adsorbed carboxylate. The size of pore can be

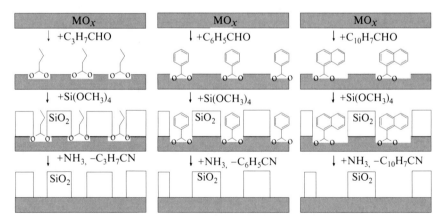

Fig. 6.4 Molecular sieving silica overlayer prepared on metal oxides using the silica CVD with butanal (*left*), benzaldehyde (*center*) and 1-naphthaldehyde (*right*) template

6.4 Fine Control of Pore-Opening Size

Table 6.1 Deposition of silica layer and its function

On the surface of	Zeolite	Metal oxide
Formed chemical bond	Si–O–Si	Si–O–M
Driving force to form the thin silica layer	Same structure (epitaxial growth)	Chemical bond between different elements (acid–base interaction)
Obtained function as catalyst and adsorbent	Pore-opening size control	Molecular imprinting
	Inactivation of external surface	Antithermal solid acid

controlled by choosing the adsorbed molecule. The idea is based on the molecular imprinting of adsorbed molecule on the metal oxide, and utilized to create the shape selective property of SnO_2 gas sensor [11]. This idea is an extension of the zeolite shape selectivity to on metal oxides.

Silica overlayer on alumina by the silica deposition is utilized to produce the antithermal support. The agglomeration of alumina into the large particle at the high temperature is strongly retarded due to the silica overlayer, as mentioned in Sect. 7.3.2 in Chap. 7.

Thus, we notice that chemical vapor deposition of silica functionalizes the metal oxide also. It is interesting that the silica is deposited on the metal oxides as well as on the zeolite as mono to ultra thin layer. However, the physical chemistry to induce the formation of think silica overlayer on the zeolite and the metal oxide is different. The formation of silica overlayer on the weak basic metal oxide is due to the interaction of the silica with the basic metal oxide. Silica is an acidic oxide, and therefore easily makes a chemical bond of Si–O–M. Therefore, the formation of silica overlayer is due to the difference between acid and base properties of silica and metal oxide.

On the other hand, the formation of silica overlayer on the zeolite is obviously different, because the silica, of which property is similar to zeolite, grows on the zeolite external surface. As mentioned below, the silica deposited on the silicalite is characterized to have a fine structure as revealed by TEM. Therefore, the formation is not induced by the chemical properties but by the similarity of the structures. Therefore, so-called the epitaxial growth of silica overlayer on the zeolite is an idea to explain the formation of overlayer with such a structure. In Table 6.1 summarized are these different profiles of the deposition of silica on the zeolite and the metal oxide.

6.4 Fine Control of Pore-Opening Size

CVD of silica to control the pore-opening size is applicable to various zeolites, MOR, MFI, LTA, BEA, MWW, etc. However, FAU Y-zeolite is only one exception, and it is impossible to control the pore-opening size of H-Y zeolite precisely.

To our best knowledge, any other unsuccessful examples are not reported to data. Its application to ultra-stable Y (USY) zeolite which is a dealuminated Y zeolite is shown later. Recently, applications to the mesoporous materials such as MCM-41 and SBA-15 are reported. However, the principle of CVD of silica does not seem to be applied to these examples of mesoporous material. Chemical liquid deposition (CLD) is also reported for a similar purpose. Discussion on these applications and proposed methods are shown later. In the present chapter, examples of the fine control based on the CVD of silica are shown to explain the separation property of mordenite, MFI, and A-zeolite.

6.4.1 Mordenite

Mordenite has the strong acidity, and applicable to various catalytic reactions. The deposition of $Si(OCH_3)_4$ on H-mordenite is therefore studied, and a high shape selectivity in the cracking of alkanes is developed as mentioned in Chap. 7. An extremely fine molecular sieving property of the silica-deposited mordenite is found. Adsorption experiments using water and xylene as the adsorbent show clearly the separation property, as shown in Fig. 6.5. Adsorption of water is unaltered by the deposition of silica, while that of o-xylene is suppressed strongly at 2.7 wt % of silica deposition. Mordenite pore size of 0.78 nm is reduced by the deposited silica, and the adsorption of large molecule (o-xylene) is retarded, but the small molecule (water) is adsorbed freely. Temperature-programmed desorption of ammonia profile is not changed by the deposition (not shown). These experimental findings clearly

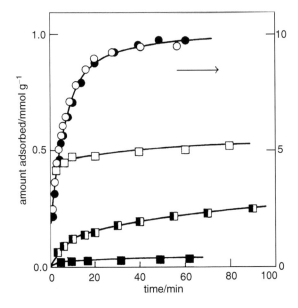

Fig. 6.5 Adsorption of water on HM (*filled circle*) and SiHM (2.7 wt %) (*open circle*), and of o-xylene on HM (*open square*), SiHM (1.4 wt %) (*half-filled square*) and SiHM (2.7 wt %) (*filled square*)

6.4 Fine Control of Pore-Opening Size

tell us that the inside of the pore is unaltered after the deposition of silica, but the pore-opening size is reduced to show the clear molecular sieving property. Thus, an important profile of the CVD technique is drawn: i.e., Si(OCH$_3$)$_4$ is deposited only on the external surface of zeolite. This is a first experimental observation to confirm the structure of silica-deposited zeolite and the molecular sieving property [12].

Adsorption of p-xylene and hexane reveals the gradual decrease of the pore-opening size by increasing the amount of silica deposited. Thus, precise control of the pore-opening size is concluded. Adsorption of a small amount of o-xylene on the SiHM with 2.7 wt % silica may take place on the external surface as estimated from the amount of deposited silica and the molecular size. Average thickness of deposited silica is found to be mono to triple layers from the external surface area and number of deposited silica. Thus, it is found that thin silica layer is enough to precisely control the pore-opening size of mordenite. Number of the silica over-layer to control the zeolite pore, however, depends on the zeolite and/or the composition of zeolite, as shown below.

6.4.2 MFI Zeolite

Two studies on the modification of MFI are shown below, since so-called HZSM-5 and silicalite with different acidic properties are studied. The HZSM-5 is an extremely important zeolite catalyst, and its applications to various catalytic reactions such as conversion of methanol to hydrocarbons, alkylation of toluene, and disproportionation of toluene are studied on the silica-deposited HZSM-5. Adsorption properties of the modified HZSM-5 are studied using o-xylene, N$_2$, and water of which sizes are so different.

Kinetics of the adsorption of o-xylene on the modified zeolites are analyzed [13] based on the equations shown below,

$$\frac{dv}{dt} = k\frac{(v_e - v)}{v}, \tag{6.6}$$

$$v_e \ln\left[\frac{v_e}{(v_e - v)}\right] - v = kt, \tag{6.7}$$

where the rate of adsorption with a constant k is described by adsorption amounts v_e and v at the equilibrium and time t, respectively, in (6.6); and integrating the adsorption amount v with respect to time t gives (6.7). Experimentally observed v and v_e are plotted due to (6.7), and the rate constant k is determined from the derived straight line. As shown in Table 6.2, the rate constant k decreases as the deposition amount increases, whereas the amount at the equilibrium v_e is kept approximately a constant. Gradual decrease of the pore-opening size, therefore, can be estimated based on the adsorption of o-xylene. On the other hand, the adsorption experiments of N$_2$ and H$_2$O show that both adsorptions change only a little, thus proving the internal surface being unaltered.

Table 6.2 Rate constant and adsorption amount at equilibrium for adsorption of o-xylene on unmodified and modified HZSM-5 [13]

Sample	$10^3\ k$ mmol min^{-1} g-ZSM^{-1}	v_e mmol g-ZSM^{-1}
HZSM-5	6.1	1.08
SiHZSM-5 (4.5 wt %)[a]	1.4	1.11
SiHZSM-5 (8.0 wt %)	1.3	1.02
SiHZSM-5 (12.0 wt %)	0.55	1.05[b]

[a]Weight gain by the CVD of silica in the parenthesis
[b]Assumed value, because it was not measured due to the slow adsorption

Silicalite is a purely siliceous zeolite with a structure of MFI. It is easy to synthesize a large crystal of silicalite so that the crystal morphology can be seen easily by SEM. A large hexagonal-flat crystal of silicalite with about 5 μm in length is used for the CVD of silicon alkoxide [14]. Because the silicalite crystal is crushed into small particles by the applied pressure in the particle preparation, the crystal is used for the CVD with a proper caution to avoid from any damages. The extent of pore-opening size enclosure is tested by the adsorption of linear and branched hexanes and butanes. Adsorption rate for butane does not change, but that of 2-methylpropane changes significantly, as the deposition degree increases, as show in Fig. 6.6. The clear and contrastive findings show the fine control of the pore-opening size of silicalite due to the deposited silica.

Thick silica layers are required for the pore-opening size control of the MFI zeolites. 15 and 23 layers of silica are deposited on the HZSM-5 with a Si/Al$_2$ ratio of 76.4 and on the silicalite, respectively, when their optimum surface concentrations of the deposited silica are compared. Relatively thick silica layer required for the modification of MFI zeolite should be a subject for characterization study.

6.4.3 A Zeolite

Separation of inorganic molecules such as O_2 and N_2 is possible by using the modified A zeolite. Inherently, these molecules can be separated using a CaNaA zeolite, commercially available as Molecular Sieve 5 A. The separation function is generated due to different adsorption properties of O_2 and N_2 on the zeolite. Because of the quadrupole moment, N_2 is adsorbed on the CaNaA in preference to O_2. In other words, a larger molecule N_2 is adsorbed preferentially, and the selective adsorption on the CaNaA zeolite is not caused by the molecular size.

Elution chromatograms of N_2, O_2, and N_2/O_2 mixture at 195 K on the NaA with and without the silica deposition are shown in Fig. 6.7 [15]. As mentioned above, N_2 is adsorbed fully, and desorption is seen little on the inherent NaA zeolite. Desorption of N_2 is observed on the deposition of silica by more than 0.41 wt %; the appearance of sharp desorption means that N_2 molecule passes through the zeolite bed without adsorption. On the other hand, O_2 is desorbed to show the peak at

6.4 Fine Control of Pore-Opening Size

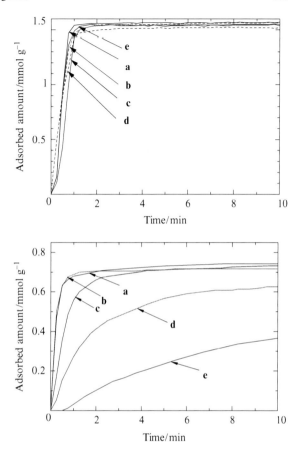

Fig. 6.6 Adsorption of butane (*upper*) and 2-methylpropane (*lower*) on silicalite (**a**), 0.75 wt % (**b**), 1.3 wt % (**c**), 1.9 wt % (**d**), and 2.7 wt % (**e**) silica deposited silicalite

2 min on the NaA. The profile of desorption changes by the deposition of more than 0.41 wt % silica. Chromatograms for a mixture of N_2/O_2 show an interesting profile; only O_2 appears on the deposition of less than 0.21 wt % silica, but on more than 0.41 wt % silica deposited, the chromatogram changes greatly and sharp desorptions of N_2 and O_2 are observed in this sequence. These chromatograms show that adsorption behaviors of molecules change dramatically due to the control of pore-opening size by the deposition of silica. N_2 is adsorbed on the NaA, but not adsorbed on the SiNaA with more than 0.41 wt % silica deposited. The pore opening-size of NaA is reduced to suppress the adsorption of N_2, while the adsorption of O_2 is not suppressed under the conditions. Molecular sizes (kinetic diameters) of N_2 and O_2 are 0.365×0.267, and $0.347 \times 0.249\,nm^2$, respectively: therefore the difference in lengths of two molecules is only as large as 0.02 nm. Separation of molecules caused by such a subtle difference in molecular sizes provides us with an interesting example of the precise control of pore-opening by the deposited silica.

Lower olefins, ethylene, propylene, and 1-butene are separated by the silica deposited NaCaA in Fig. 6.8 [16]. These olefins are adsorbed on the NaCaA in

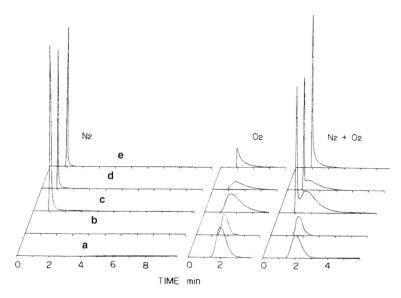

Fig. 6.7 Elution chromatograms of N_2, O_2, and N_2/O_2 mixture at 195 K on the NaA (**a**) and SiNaA with the silica deposition of 0.09 (**b**), 0.21 (**c**), 0.41 (**d**), and 0.55 (**e**) wt%. Reproduced from [15]. Copyright 1991 American Chemical Society

a similar manner. The deposition of silica suppresses the adsorption of olefins, 1-butene, propylene, and ethylene in this sequence of molecular size. Adsorption of 1-butene is suppressed first by the silica deposition, and adsorption of propylene is affected by the further deposition. Finally at the deposition of 1.30 wt % silica, only ethylene is adsorbed. The separation of these olefins on the SiNaCaA also is an example of the fine control of the pore-opening size of A zeolite. It is anticipated that these modified zeolites are applied to the process of separation industrially, because the material with the novel functions may provide us with a chance to develop a new process of separation.

CVD of silica on NaA zeolite is studied by a continuous-flow method [17]. The aim of the study is to develop the method for producing a large amount of CVD zeolite which could be utilized in the industrial scale. Zeolite NaA sample in a Pyrex-glass tube reactor (12 mm i.d.) is separated vertically into three portions, i.e., top, middle, and bottom beds. About 3 g of the zeolite is contained in each portion and the total length of the sample bed is about 7 cm. Silica is deposited at 473–673 K, and the proceeding of the deposition is monitored by the undeposited portion of the alkoxide. The deposition of silica is saturated at 573 K to form a silica layer of which thickness is about mono to triple; however, it is activated by admission of water to continue the deposition. Pretreatment by flowing He gas at 673 K affects strongly the deposition; and it may be suspected that water remained in the NaA may control the deposition of silica. Under the studied experimental conditions, the pretreatment for 20 h is required to produce the SiNaA which is modified homogeneously along the sample bed.

6.4 Fine Control of Pore-Opening Size

Fig. 6.8 Adsorption of ethylene, propylene, and 1-butene at 273 K on NaCaA (**a**) and SiNaCaA with 0.42 (**b**), 0.64 (**c**), 0.88 (**d**), 1.18 (**e**) and 1.30 (**f**) wt% silica deposition

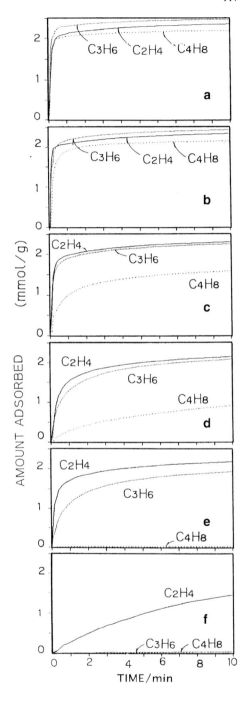

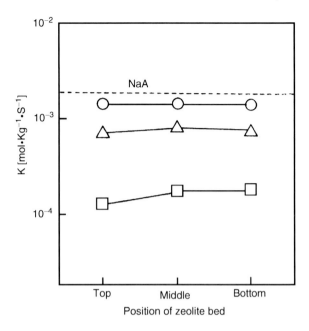

Fig. 6.9 Axial distribution of k for nitrogen adsorption along the sample bed on SiNaA prepared by the repeated deposition; times 1 (*open circle*), 2 (*open triangle*), and 4 (*open square*). *Dotted line* shows the k for NaA. Reproduced with permission from [17]. Copyright 1990 The Society of Chemical Engineers, Japan

Fig. 6.10 Relation between k for nitrogen adsorption and n, Si/Al ratio measured by XPS on NaA (*filled circle*) and SiNaA modified at 473 (*open circle*), 573 (*open triangle*), and 673 (*open square*) K. Only one sample marked by (*filled square*) was prepared by the vacuum equipment of silica CVD

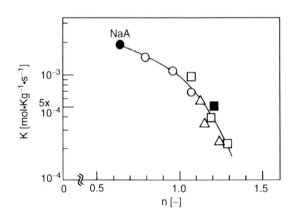

Figure 6.9 shows the thus formed axial distribution of rate constant k for nitrogen adsorption along the zeolite bed which is derived from (6.7). The value of k is almost constant through the sample beds, and it decreases with increasing the deposition time. Figure 6.10 shows the relation of the value of k with the Si/Al ratio measured by XPS, when the sample is produced at different temperatures and with repeated cycles of deposition. Interestingly, the value of k depends only on the n value, i.e.,

the amount of silica deposited, but neither on the method of CVD (continuous-flow or vacuum) nor on the deposition temperature. This study shows that the continuous-flow method can be applied to produce the CVD zeolite, and the utilization depends on careful control of the experimental conditions.

6.4.4 Y Zeolite

Y zeolite shows the unusually behavior, when the alkoxide is deposited. Silicon alkoxide can enter the pore of Y zeolite, because the amount of deposited silica is much larger than expected for the deposition on the external surface. Unmodified Y zeolite with 5.6 of Si/Al_2 ratio shows the amount of adsorbed nitrogen, 207 cm^3 g^{-1}, while it decreases to only 2.9 cm^3 g^{-1} upon the deposition of silica by 14 wt % at 593 K. The deposition of silica takes place inside the pore of Y zeolite, and modifies the property greatly. The principle mentioned above for various zeolites therefore cannot be applied to the H-Y zeolite [18].

However, the high silica Y zeolite, which has the Si/Al_2 ratio 15–40, is modified by the CVD to control the pore-opening size. Figure 6.11 shows the uptake of 1,3,5-triisopropylbenzene on the unmodified and modified Y zeolites; the adsorbed amount decreases gradually by increasing the amount of deposited silica. Amount of N_2 adsorbed on these samples does not change largely. Therefore, the deposition of silica seems to take place on the pore-entrance to finely control the size of pore-opening. The high silica H-Y zeolite is known as the ultra stable Y (USY) zeolite with the strong solid acidity. Thus, the deposition occurs rapidly on the zeolite site close to the pore-entrance. The deposited silica which is either on the external surface or on the pore-entrance controls the pore-opening size.

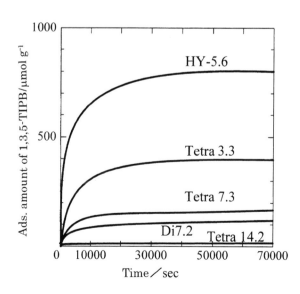

Fig. 6.11 Uptake of 1,3,5-triisopropylbenzene on the unmodified (H-Y-5.6) and modified Y zeolites using tetra- and di-methoxysilane (Tetra and Di, respectively) with different amounts of silica shown by the number

Another way to overcome the problem of H-Y zeolite is the deposition on the H-Y zeolite filled with the organic molecules. Diethylether is adsorbed on the H-Y zeolite, followed by deposition of silica. Because the pore is filled with the ether, silica is deposited only on the external surface. Using this method, the pore-opening size is controlled, and the adsorption of larger molecules is suppressed gradually with increasing the amount of silica deposited. CVD of $Si(OC_2H_5)_4$ on the Y zeolite with the ether-filled pores is thus reported by Itoh et al. as a useful application to Y-zeolite [19].

6.5 Characterization of Deposited Oxide

6.5.1 XPS Measurements

X-ray photoelectron spectroscopy (XPS) on the H-mordenite reveals Si_{2s} and Al_{2p} signals at 161 and 82 eV, respectively, and the intensity ratio corrected by cross sections is consistent with the constituent ratio of the zeolite, thus justifying the measurements of XPS. As the deposition of silica increases, the intensity of Al_{2p} signal decreases. Thus measured Si/Al ratio is larger than the ratio expected when the silica is deposited nonselectively not only on the external surface but also inside the zeolite pore, as shown in Fig. 6.12a. The measured ratio higher than the value of total composition shows that silica is deposited on the external surface. From these measurements of XPS, the thickness of silica overlayer is estimated. When

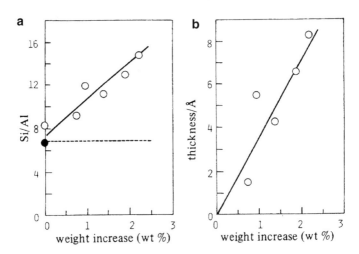

Fig. 6.12 Si/Al ratio measured by XPS on the mordenites (**a**); and the thickness of silica derived from (5.8) (**b**). The *dotted line* in (**a**) shows the change of Si/Al when the silicon atom is deposited nonselectively not only on the external but also in the internal of zeolites. Reproduced with permission from [12]. Copyright 1984 Royal Society of Chemistry

6.5 Characterization of Deposited Oxide   121

the escape depth of these electrons is assumed to be 2.0 nm, XPS shows the compositions of Si and Al in the silica overlayer and the zeolite, sum of which thickness is 2.0 nm. Thus, the Si/Al ratio measured by XPS is described as

$$\frac{\text{Si}}{\text{Al}} = \frac{(1+n) + (2-t)\frac{n}{1+n}}{(2-t)\frac{1}{1+n}}, \quad (6.8)$$

where n is the Si/Al ratio measured in the unmodified zeolite. The thickness t is measured from the XPS derived intensity based on (6.8), as shown in Fig. 6.12b. It is found that the thickness is about 0.3 nm in the deposition of 1.0 wt % SiO_2. The averaged surface concentration of silica at this weight gain is 12.2 Si nm^{-2} from the external surface area (3.6 m^2 g^{-1}). Because the T-site surface concentration is 8.6 nm^{-2}, this concentration corresponds to the mono to double silica overlayer. Because the length of Si–O bond is ca. 0.2 nm, the measured thickness also suggests the formation of mono to double layers. Thus, the XPS measurement of the deposited silica supports the principle for the CVD of silicon alkoxide that it is deposited only on the external surface of the zeolite to form the very thin layer.

6.5.2 *EXAFS of the Deposited Germanium*

In place of Si(OCH$_3$)$_4$, Ge(OCH$_3$)$_4$ is deposited on the H-mordenite to measure the structure of deposited oxide by means of an EXAFS (extended X-ray absorption fine structure) [5]. Ge and Si are atoms with the same electron configuration in the outer orbital, and both alkoxides have the same structure. Fourier transforms of the EXAFS spectra exhibit two peaks at ca. 0.14 and 0.27 nm ascribed to Ge–O and Ge–Ge, respectively, as shown in Fig. 6.13 and Table 6.3. The distances of Ge–O and Ge–Ge are 0.173 and 0.313 nm, respectively, which are in good agreement with those of α-quartz type GeO$_2$. Coordination numbers by O and Ge with the centered Ge atom are 4.0 and 1.1, respectively. Because the coordination numbers of Ge with O and Ge atoms in α-GeO$_2$ are 4.0 and 4.0, respectively, the EXAFS of the deposited oxide shows that the deposited Ge has the structure similar to α-GeO$_2$ in a tetrahedral configuration, but it is not a large particle, but dispersed highly on the surface, or stabilized in the extremely thin layer. Experiments of the cracking of alkane isomers show the enhanced shape selectivity, when the deposited oxide keeps the structure of thin oxide layer. Therefore, it is concluded that the thin metal oxide layer of germanium oxide controls the pore-opening size precisely.

The germanium oxide deposited H-mordenite is unstable in the humidified conditions. When it is stored in the humidified conditions, and then used for the catalytic cracking, the shape selectivity is not observed at all. The deteriorated sample of the Ge–HM reveals the increase of the coordination number of Ge–Ge. This means that the metal oxide is agglomerated into a large particle of GeO$_2$ which does not control the pore-opening size. In other words, the thin metal oxide layer is required for the

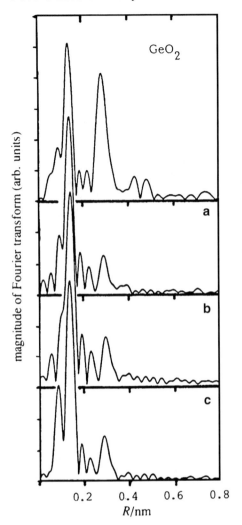

Fig. 6.13 Fourier transforms of the EXAFS spectra measured on GeO$_2$ and Ge deposited mordenite by the surface concentration of 10.0 (**a**), 16.0 (**b**) and 30.2 (**c**) Ge nm^{-2}. Reproduced with permission from [5]. Copyright 1984 Royal Society of Chemistry

Table 6.3 Ge deposited on HM and in a bulk of GeO$_2$ characterized by an EXAFS measurement

Sample	CVD (time min^{-1})	Deposited Ge/nm^{-2}	Distance/Å Ge–O	Ge–Ge	Coordination number Ge–O	Ge–Ge
GeHM10	10	10.0	1.73	3.13	4.9	1.1
	60	16.0	1.73	3.14	4.6	1.1
	1,080	30.2	1.73	3.13	4.9	1.2
GeO$_2$	0	0	1.73	3.15	4.0	4.0

6.5 Characterization of Deposited Oxide

selectivity control. GeO$_2$ is dissolved in water unlike as SiO$_2$, and therefore readily changes the structure. This is the reason why the GeO$_2$ deposited sample is unstable to the environmental humidity.

6.5.3 TEM Observation

Transmission electron microscopy (TEM) observation is difficult in usual zeolite samples, because it is hard to discriminate between the deposited layer and the zeolite. However, clear TEM images are observed in the silica deposited on a silicalite crystal, because it may be a large and fine crystal.

The silica layer deposited by 1.9 wt % on the external surface of the silicalite crystal is seen clearly in Fig. 6.14 [20]. Silicalite is coated uniformly with the silica

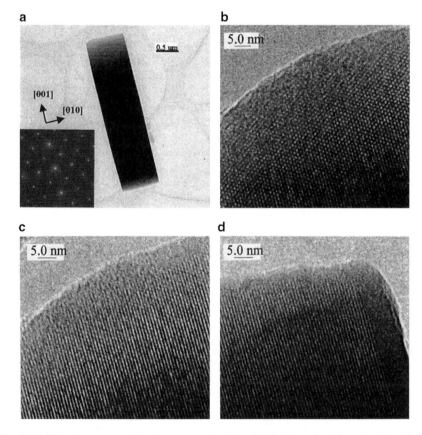

Fig. 6.14 TEM observation of the silicalite on the 1.9 wt % silica deposited along the (100) direction with diffraction pattern inset (**a**). (**b**) Enlarged portion of *upper left corner* of crystal in (**a**). (**c**) Same portion as in (**b**) but with slightly different focus. (**d**) Enlarged section of *upper right corner* of crystal in (**a**)

layer of a thickness of about 5–8 nm, and the measured thickness agrees well with the value 5.5 nm estimated from the amount of silica and the external surface. Relatively thick silica overlayer is required to control the pore-opening size of silicalite, and thus prepared silicalite shows the fine selectivity of adsorptions of linear alkane, as shown in Fig. 6.6. It is extremely interesting that such linear alkanes as hexane and butane are adsorbed freely on the modified zeolites. In other words, hexane and butane pass through the deposited silica layer to be adsorbed on the silicalite.

In addition, the silica layer shows the interesting profile of TEM observations. Pictures of (b) and (c) are taken at the same place of left corner of the crystal but with different focus. An amorphous like oxide covers the zeolite in (b), while the fringes straightly arrive at the outer surface in (c). The TEM image of the silica deposited changes by adjusting the degree of focus: i.e., an amorphous like image of silica changes into a clear crystal like pattern, and the fringes running in the crystal arrive at the outermost of the sample. In the picture (d), the fringes are seen straightly in the direction of (001), but the amorphous oxide is formed in the direction of (010).

This feature of the TEM image suggests that the deposited silica is formed as the material which is neither a pure crystal nor a mere amorphous oxide. It may be therefore concluded that the silica layer is elongated on the top of the zeolite crystal along the pore direction. Silica is loaded on the pure silicalite, and therefore the nearly epitaxial formation of silica layer may be suggested. Most probably, the formed silica has a structure, which is affected by the structure and composition of the basal plane of zeolite. One may have an idea that a difference between the structures of the deposited silica and the top of zeolite controls the pore-opening size.

6.6 External Surface Acidity: Measurements and Inactivation

The acid sites on the external surface are regarded as the nonselective site for the shape selective catalytic reaction, and therefore these are removed or inactivated. The deposition of silica inactivates the external surface acidity, thus enhancing the shape selective reaction; therefore the inactivation or passivation by the deposited silica is another reason for the enhancement of shape selectivity due to the CVD of silica. However, the external surface acidity is not sufficiently characterized, because it is not studied easily.

Cracking of 1,3,5-triisopropylbenzene is used to test the external surface acidity; this molecule is difficult to enter into the pores of MFI and MOR because of its molecular size. The decrease of the cracking activity due to the deposition of silica on the mordenite with the different compositions of Al and cation is shown in Fig. 6.15 [21]. In this experiment, the conversion level of the parent zeolite is adjusted to 15–20% by controlling the sample weight, and the modified Na-type mordenite (NaM20) is protonated before the cracking reaction. It is interesting that the degree in decrease of the cracking activity by the silica deposition is independent of the molar ratio of Si/Al_2 10 and 20, cations H (HM) and Na (NaM), and

6.6 External Surface Acidity: Measurements and Inactivation

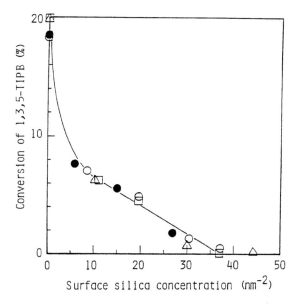

Fig. 6.15 Inactivation of the external surface by the deposition of $Si(OCH_3)_4$ on HM-10 (*open square*), HM-20 (*open circle*), NaM-20 (*filled circle*), and DHM-20 (dealuminated) (*open triangle*) mordenites. Reproduced with permission from [21]. Copyright 1993 Elsevier

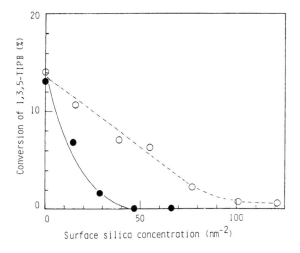

Fig. 6.16 Inactivation of the external surface by the deposition of $Si(OCH_3)_4$ on HZSM-5-1 (*open circle*) and -2 (*filled circle*). Reproduced with permission from [21]. Copyright 1993 Elsevier

before and after the dealumination. On the other hand, two kinds of HZSM-5-1 and -2, show clearly different behaviors of the inactivation, as shown in Fig. 6.16. Characterization of these zeolites by means of NMR (nuclear magnetic resonance) and XPS shows that the HZSM-5-1, on which a larger surface concentration of

Table 6.4 Brønsted acidity of the unmodified and modified HZSM-5, and the acidity on the external surface derived from the difference between them

Sample	Number of the acid site per mol kg^{-1}	Strength of the acid site per kg mol^{-1}	Surface area per m^2 g^{-1}	Surface density of the acid site per nm^{-2}
HZSM-5[a]	0.361	140	372	0.58
SiHZSM-5[b]	0.251	142	–	–
Difference[c]	0.110	146	26.6[d]	2.49

[a] Prepared from JRC-Z5–90Na by an exchange with NH$_4$NO$_3$, followed by calcination
[b] Modified by the CVD of 11.6 wt % silica
[c] Difference between unmodified and modified HZSM-5 samples
[d] External surface area of HZSM-5 measured by the benzene-filled pore method

deposited silica is required for the inactivation, has not only the large amount of extra framework Al but also the aluminum enrichment on the external surface.

An effect of attached alkyl group to Si upon the inactivation of the external surface is tested using Si(OR)$_x$R$_{1-x}$, where $x = 1, 2, 3$, and 4 and R = CH$_3$ and C$_3$H$_7$ [4]. Monomethoxy silane is found to be effective in the inactivation, and in particular monomethoxy tripropyl silane is the most effective; and the monolayer of silica (9.3 Si nm^{-2}) is enough to totally inactivate the external surface acidity.

Cracking of alkylbenzene, e.g., cumene, is believed to take place on the Brønsted acid sites. Therefore, these studies show the inactivation of Brønsted acid sites due to the deposited silica. The HZSM-5, JRC-Z5–90H, provided by the Catalysis Society of Japan as a reference catalyst has the unusual high concentration of the external surface Brønsted acidity [22]. The Brønsted acidity is characterized by utilizing the inactivation of the large amount of acid sites.

Acidic properties of the HZSM-5 samples unmodified and totally inactivated by silica of 11.6 wt % and the surface concentration 44 Si nm^{-2} are measured by the ammonia TPD, as shown in Table 6.4. The difference between them seems to give us the profile of Brønsted acidity on the external surface, and it is about one thirds of the total acid sites. The strength of the external Brønsted acidity is a small degree stronger than usual, and the surface density of the acid site is much larger than those of inner acid sites of HZSM-5, as shown by Table 6.4.

Brønsted acid site is generated with a fine structure of the crystal, on which the strength depends, as mentioned in Chaps. 2–4. Therefore, Brønsted acid sites, when located anyplace either on external or internal, should have the same structure and acid strength. On this sample, Al concentration is made rich on the external surface of small particles, and the surface density of the acid site is enhanced greatly, more than expected from the usual HZSM-5. Strength (146 kJ mol^{-1}) and surface density (2.49 nm^{-2}) of the Brønsted acid sites on the external surface thus measured are interesting profiles of the external surface acidity; this is the first report of the quantification of external acidity on zeolite.

References

1. S.M. Csicsery, Pure Appl. Chem. **58**, 841 (1986)
2. M. Niwa, H. Itoh, S. Kato, T. Hattori, Y. Murakami J. Chem. Soc. Chem. Commun. 819 (1982)
3. N.R.E. Impens, P. van der Voort, E.F. Vansant, Microporous Mesoporous Mater. **28**, 217 (1999)
4. J.H. Kim, A. Ishida, M. Okajima, M. Niwa, J. Catal. **161**, 387 (1996)
5. T. Hibino, M. Niwa, Y. Murakami, M. Sano, J. Chem. Soc. Faraday I 85, 2327 (1989)
6. T. Hibino, M. Niwa, Y. Murakami, M. Sano, S. Komai, T. Hanaichi, J. Phys. Chem. **93**, 7847 (1989)
7. M. Inomata, M. Yamada, S. Okada, M. Niwa, Y. Murakami, J. Catal. **100**, 264 (1986)
8. M. Niwa, Y. Kawashima, T. Hibino, Y. Murakami, J. Chem. Soc. Faraday I **84**, 4327 (1988)
9. T. Hibino, M. Niwa, A. Hattori, Y. Murakami, Appl. Catal. **44**, 95 (1988)
10. N. Kodakari, N. Katada, M. Niwa, J. Chem. Soc. Chem. Commun. 623 (1995)
11. T. Tanimura, N. Katada, M. Niwa, Langmuir **16**, 3858 (2000)
12. M. Niwa, S. Kato, T. Hattori, Y. Murakami, J. Chem. Soc. Faraday Trans. I **80**, 3135 (1984)
13. M. Niwa, M. Kato, T. Hattori, Y. Murakami, J. Phys. Chem. **90**, 6233 (1986)
14. H.A. Begum, N. Katada, M. Niwa, Microporous Mesoporous Mater. **46**, 13 (2001)
15. M. Niwa, K. Yamazaki, Y. Murakami, Ind. Eng. Chem. Res. **29**, 38 (1991)
16. M. Niwa, K. Yamazaki, Y. Murakami, Ind. Eng. Chem. Res. **33**, 371 (1994)
17. K. Yamazaki, M. Niwa, Y. Murakami, Kagaku Kougaku Ronbunshu **16**, 564 (1990)
18. J.H. Kim, Y. Ikoma, M. Niwa, Microporous Mesoporous Mater. **32**, 37 (1999)
19. H. Itoh, S. Okamoto, A. Furuta, Nippon Kagaku Kai-shi 420 (1989)
20. D. Lu, J.N. Kondo, K. Domen, H.A. Begum, M. Niwa, J. Phys. Chem. B. **108**, 2295 (2004)
21. T. Hibino, M. Niwa, Y. Murakami, Zeolites **13**, 518 (1993)
22. K. Tominaga, S. Maruoka, M. Gotoh, N. Katada, M. Niwa, Microporous Mesoporous Mater. **117**, 523 (2009)

Chapter 7
Application of the CVD of Silica to the Shape Selective Reaction

Abstract Chemical Vapor Deposition (CVD) of silica is applied to achieve the fine shape selectivity on mordenite, ZSM-5, and A-zeolite is mentioned. Selective *para*-dialkylbenzene formation due to the alkylation and the disproportionation is the most important and interesting reaction achieved on the silica-deposited H-ZSM-5. Various industrial applications of the CVD of silica are reviewed.

7.1 Selective Formation of *Para*-Dialkylbenzene

7.1.1 Principle of the Shape Selectivity

How is the high selectivity to produce *p*-xylene realized on the zeolite HZSM-5 synthesized by a usual method? To reply the question, zeolite ZSM-5 samples with ca. 75 ± 20 of Si/Al_2 ratio are synthesized under different conditions; these zeolites are comprehensively characterized, and the relation of the characterized parameter with the obtained selectivity is studied.

ZSM-5 samples with various morphologies are obtained, and the sizes of the synthesized zeolites are found to be 0.5–6 μm from the SEM image. Using these samples of HZSM-5, a principle of the shape selective formation of *p*-xylene as a result of alkylation of toluene with methanol is examined based on characterization data of the adsorption rate and the external acidity [1]. Selectivity to *p*-xylene formation at the conditions of 20% conversion of toluene is measured and plotted against the diffusion rate constant of adsorption of *o*-xylene in Fig. 7.1. Because selectivity depends on the conversion level, measurements of the selectivity at a constant conversion are required in order to examine its dependence on the characterization parameter. The selectivity at 20% of the conversion is clearly correlated with the rate constant of *o*-xylene adsorption; it increases with decreasing the diffusion rate constant except for one sample named No. 15. The rate constant of adsorption is simply related with the crystal size of HZSM-5 in Fig. 7.2, and it is supported theoretically by the equation shown below.

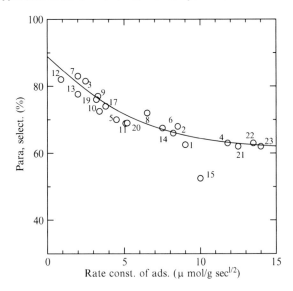

Fig. 7.1 Selectivity to *p*-xylene formation at the conditions of 20% conversion of toluene plotted against the diffusion rate constant of adsorption of *o*-xylene. Numbers denote the HZSM-5 synthesized at different conditions

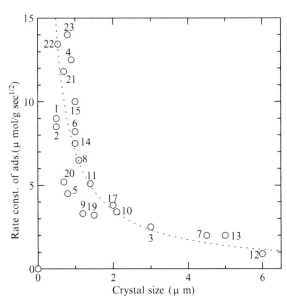

Fig. 7.2 Dependence of the adsorption rate constant of *o*-xylene upon the crystal size of HZSM-5

$$k = M_e \left(\frac{6}{r}\right)\left(\frac{D}{\pi}\right)^{0.5}, \qquad (7.1)$$

where k is the rate constant of adsorption and r, D, and M_e are radius of the zeolite crystal, diffusion constant, and adsorption amount at equilibrium, respectively.

7.1 Selective Formation of *Para*-Dialkylbenzene   131

The adsorption rate of *p*-xylene can be regarded to change little in these zeolites, and therefore the selectivity increases with increasing the difference in the rates of adsorption of *p*- and *o*, *m*-xylenes. With increasing the difference (or the ratio) in adsorption rates of *p*- and *o*, *m*-xylenes, and also with increasing the crystal size, the selectivity to *p*-xylene increases. A large deviation from the relationship observed in a sample No. 15 is due to the unusual high concentration of external surface acid site; an inhomogeneous distribution of the acid site on this sample is implied from the experimental finding of the abundance of dislodged Al. We learn from these experimental studies that the selectivity is mainly controlled by the difference in rates of adsorption of molecules, and the unusual high external surface acidity results in the low selectivity. In order to enhance the selectivity, one has to notice that the rate of adsorption is controlled principally, and the activity on the external surface also is taken into consideration.

7.1.2 CVD Zeolite to Produce the Para-Dialkylbenzene

As shown in Chap. 6, Niwa et al. found that the deposition of silica on the external surface of zeolite improved the shape selectivity of zeolite. Selective formation of *para*-dialkylbenzene has been studied extensively based on the silica deposited zeolite, and much attention has been paid to the enhanced high selectivity. Production of *p*-xylene and *p*-diethylbenzene due to the alkylation and the disproportionation has been studied as the industrial application process.

The high selectivity to form *p*-xylene was found in a study of methanol conversion into gasoline on the ZSM-5 modified with CVD of silica in 1986 [2]. With increasing the amount of deposited silica, the yield of products shifts to small hydrocarbons. Of much more interest is the change of yield distribution of xylene isomers by the deposited silica. The distribution of xylene isomers on the HZSM-5 is about 1:2:1 for *o*:*m*:*p* xylenes, which is in agreement with the thermodynamics-controlled products distribution at ca. 623 K. Deposition of silica decreases *o*- and *m*-xylene yields, and in place of these larger xylene isomers, *p*-xylene yield is increased. The selectivity to form *p*-xylene increases gradually and finally becomes nearly 100% at about 13 wt% of silica deposition as shown in Fig. 7.3.

Change of the xylene isomers yield due to the deposited silica is later confirmed by the study on alkylation of toluene with methanol to *p*-xylene formation, which is one of the most important industrialized reactions [3]. The same dependence of the *p*-xylene selective formation upon the deposition amount of silica is observed also in the toluene disproportionation and *o*-xylene isomerization; thus it is suggested that the selectivity is controlled by a common parameter existing in these reaction steps. Obviously, the difference in diffusion rates of xylene isomers determines the selectivity to form *p*-xylene. This is the most important example of the product shape selectivity, successfully realized by the deposition of silica on the external surface of HZSM-5. The HZSM-5 studied has 21 of the Si/Al ratio and 0.25 µm

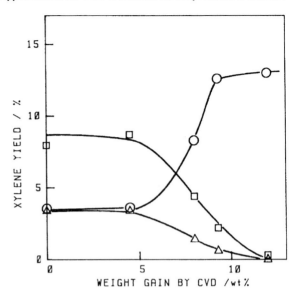

Fig. 7.3 Change of the products xylene isomers yield with the silica deposition; o- (*open triangle*), m- (*open square*), and p- (*open circle*) xylenes

of the crystal length. Because the external surface area is 11 m² g⁻¹, and the highest selectivity is realized at the deposition of 121 Si nm⁻² of silica, which is much larger than those confirmed recently on a different HZSM-5 (ca. 40 Si nm⁻²) [4]. Figure 7.4 shows a model of silica-modified MFI zeolite. In the micropore, o- and m-xylenes are possible to be formed, whereas only p-xylene can pass through the pore mouth which is narrowed by the silica overlayer.

The dealkylation and realkylation (i.e., disproportionation) of ethylbenzene to form diethylbenzene is investigated over the pore size-regulated MFI zeolite by the deposition of silicon tetraethoxide [5]. Halgeri et al. belonging to Indian Petrochemical Corporation report an excellent achievement of the selective production of p-diethylbenzene, as shown in Table 7.1. The selectivity to form the *para* isomer is enhanced up to about 100% with the conversion of ethylbenzene kept almost a constant, when the reaction conditions such as temperature, WHSV, and H₂/hydrocarbon ratio are optimized. They have industrialized the p-diethylbenzene production using the pore size-regulated MFI zeolite. With this development, India became the third country in the world to develop and commercialize the eco-friendly technology for p-diethylbenzene manufacture, according to Halgeri. The two others are UOP, USA and TSMS, Taiwan, and at least two of the three companies utilize the CVD zeolite, most probably.

Wang et al. report an excellent selectivity of the formation of p-ethyltoluene as a result of toluene ethylation with ethylene on the silica deposited HZSM-5 [6]. The composition of isomers at the equilibrium conditions at 600 K, p:m:o ethyltoluenes = 33.7:49.9:16.5, dramatically changes, and 99.4% of the selectivity to p-ethyltoluene is achieved on the silica deposited HZSM-5. Priority of the CVD of silica in preference to the impregnation of metal oxides is confirmed in their studies.

7.1 Selective Formation of *Para*-Dialkylbenzene

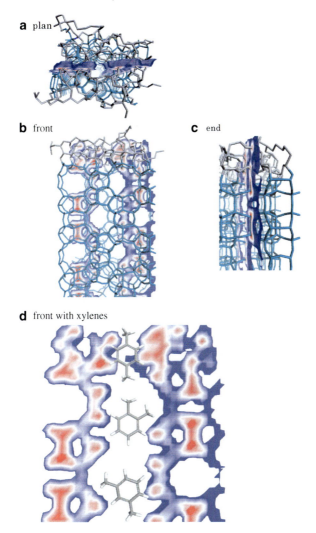

Fig. 7.4 A model of silica-modified MFI zeolite. (**a–c**) Orthographic drawing (third angle projection) of frameworks of MFI (*blue*) and silica overlayer (*gray*). A cross-section of micropore is shown by a plane in which the zeolite framework is shown by *red*. (**d**) The cross-section with xylene isomers

Inactivation of the external surface and narrowing of the pore-opening size are made clear in these studies, and both are believed to contribute to enhancing the shape selectivity. It is interesting, however, to study which change of the properties enhances the *para*-isomer selectivity effectively. In the study of the CVD of silica using different silane compounds, the modified zeolites are precisely characterized to understand these effects on the degree of selectivity enhancement [7]. *o*-Xylene adsorption and cracking of 1,3,5-triisopropylbenzene are measured for

Table 7.1 Selective production of p-diethylbenzene

	Catalyst	
	MFI	Silylated MFI
Ethylbenzene conversion (%)	12.83	12.5
Selectivity to (%)		
Benzene	41.31	38.72
Diethylbenzene	56.66	54.32
Others	2.03	6.56
Diethylbenzene composition (%)		
Para	33.43	99.42
Meta	63.96	0.58
Ortho	2.61	–

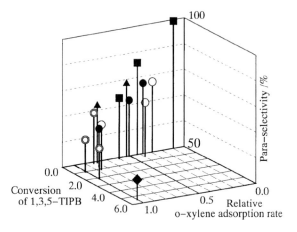

Fig. 7.5 Selectivity of p-xylene formation from the alkylation of toluene plotted against the relative o-xylene adsorption rate and the degree of cracking of 1,3,5-triisopropylbenzene. Symbols show the samples ZSM-5 unmodified (*filled diamond*) and modified by different silicon alkoxide compounds; $Si(OCH_3)(C_3H_7)_3$, (*double circle*); $Si(OCH_3)(CH_3)_3$, (*filled circle*); $Si(OCH_3)_2(CH_3)_2$, (*open circle*); $Si(OCH_3)_3(CH_3)$, (*filled triangle*); $Si(OCH_3)_4$, (*filled square*)

characterizations of controlled degrees of the pore-opening size and the external surface acidity, respectively. The selectivity of p-xylene formation from the alkylation of toluene measured at a constant conversion level of toluene is plotted against both parameters in Fig. 7.5. The selectivity plotted vertically in Z-axis first increases with inactivation of the external surface. However, it increases up to ca. 70% at most. To have the selectivity of more than 90%, the pore-opening size is required to be controlled, as confirmed in the figure. This means that the inactivation of the external surface is insufficient and the pore-opening control results in the high selectivity which is close to almost 100%.

O'Connor et al. report the repeated cycle process of the silica CVD to enhance the selectivity to p-xylene in the disproportionation of toluene [8]. One cycle of the deposition in a continuous flow method consists of (1) the deposition of tetraethoxysilane at 373 K, (2) flushing with nitrogen, and (3) calcination by air at 773 K. Selectivity to form *p*-xylene increases gradually, and 16 times of the repeated cycle are required to achieve the excellent selectivity, as high as 100%. The slow and stepwise CVD technique may be carried out to take care of the homogeneous and precise modification, and this is an example to obtain the extremely high selective zeolite by means of the silica deposition technique. *p*-Xylene distribution increases steeply from the ninth cycle, on which the external surface has been inactivated almost sufficiently. The finding is similar to the observation stated above, and the inactivation of the external surface does not fully enhance the shape selectivity of the *para* isomer production.

7.1.3 In Situ Production of CVD Zeolites

Various methods of the silica CVD are proposed to produce the active, selective, and durable zeolite catalysts. Among them, in situ production is known as the method applicable to the industrial scale production of the CVD zeolite. This method is first reported by Wang et al. [9]; they use a mixture of 50% toluene, 45% methanol, and 5% $Si(OC_2H_5)_4$, which is flowed to the zeolite with a carrier gas N_2. The selectivity of the product *p*-xylene is monitored to identify the degree of modification. The CVD is performed in situ in a reactor of the toluene alkylation. Temperature of the deposition is ca. 473 K, and the degree of deposition is measured from the amount of unconverted alkoxide. Occasionally, H_2 and/or H_2O are admitted simultaneously together with the alkoxide to control the deposition reaction. In situ monitoring of the deposition degree using the test reaction of the *p*-xylene formation is useful to prepare the CVD zeolite in a large scale. Water is produced as a reaction of methanol with toluene, and the water produced maybe activates the zeolite surface to continue the deposition of silica. It may be effective in homogeneous modification by the deposited silica.

In a method proposed by Halgeri et al., H_2 is used as a carrier gas, and the mixture of 6.5% tetraethoxysilane in toluene and methanol is contacted with the zeolite at 503 K for the desired hours [10]. After hydrogen is switched to nitrogen, the temperature of catalyst bed is elevated to 815 K, and the modified zeolite is calcined by oxygen for 10 h to obtain the silica deposited ZSM-5. These total optimizations, i.e., the mixture of a small amount of tetraethoxysilane in toluene and methanol, the deposition temperature at 500 K, the carrier gas of hydrogen and the calcination step at a high temperature may be important to produce the homogenously modified zeolite. This method is proposed by the company, which successfully utilizes the CVD method for the industrial process, and therefore could be recommended for a method of production of large amounts of modified zeolite. Scientific bases for the preparation of CVD zeolites, i.e., the mechanism of preparation and effects of the methods upon the zeolite function, will be subjects of the investigation carried out in near future.

7.1.4 HZSM-5 In Situ and Ex Situ Prepared for the CVD of Silicon Alkoxide

The zeolite acidity is lost to some degree, when it is exposed to the atmosphere humidity. On the other hand, the H-zeolite prepared in situ in the reactor and unexposed to the atmosphere shows the fine solid acidity. The difference in the solid acidities between in situ and ex situ prepared H-zeolites becomes outstanding at the conditions of high concentration of acid site, as mentioned in Chap. 2 [11]. From the findings of the solid acidity, it may be expected that the external surface also changes when it is exposed to the humidity. A characterization study on the in situ and ex situ prepared HZSM-5s shows that the external surface area increases, and the external surface acidity decreases, when the HZSM-5 is prepared ex situ in compared with the in situ prepared one [4]. Therefore, we must notice that the external surface also changes by the humidity to some degree.

One can compare between the degrees of enhancement of p-xylene formation by the deposition of silica on the in situ and ex situ prepared HZSM-5, as shown in Fig. 7.6. The selectivity to form p-xylene increases linearly with increasing the deposited silica on four kinds of in situ prepared HZSM-5 in Fig. 7.6a. On the other hand, the selectivity increases nonlinearly on the ex situ prepared HZSM-5s in Fig. 7.6b, and it increases with increasing the amount of deposited silica in different manners, which depends on the sample. In situ prepared HZSM-5 is therefore regarded as the starting material of the HZSM-5 with a preferential property, because the increment of the selectivity depends linearly on the amount of silica deposited, and therefore it is easy to design the selective catalyst from the amount of silica deposited.

As discussed in the TEM measurements in Chap. 6, the silica deposited on the external surface is not a simple amorphous material, but the structure is regulated by the basal plane of zeolite. Therefore, it is expected that the in situ prepared conditions of the HZSM-5 provide us with the external surface adequate for the fine structure of the deposited silica. The external surface with such an unaltered fine structure seems to stabilize the deposited silica to control the pore-opening size, of which degree depends linearly on the amount of silica.

7.1.5 CLD of Silica for the Shape Selective Adsorption

Performance of the materials modified by CVD depends principally on the temperature and gaseous composition during the deposition reaction. This can be an advantage compared to the conventional liquid phase methods such as impregnation and gelation, because the control of temperature and gaseous composition is easy for the manufacturers who have operated plants for catalytic reactions in vapor phase. In liquid phase methods, the temperature is complicatedly affected by many factors, and hence the control of temperature is difficult especially during the elevation

7.1 Selective Formation of *Para*-Dialkylbenzene 137

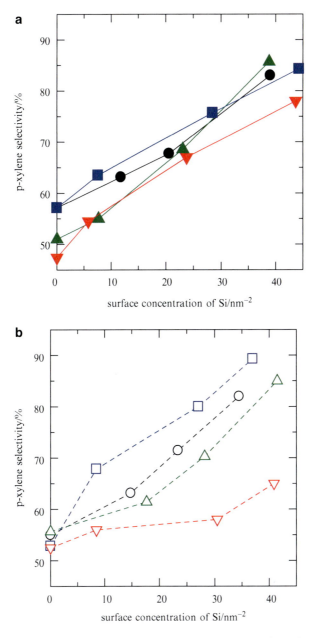

Fig. 7.6 Comparison between the degrees of enhancement of *p*-xylene formation by the deposition of silica on the in situ (**a**) and ex situ (**b**) prepared HZSM-5. Each four kinds of HZSM-5 shown by different notations are tested

of temperature. This sometimes gives different results in different scale of liquid phase production of functional materials. On the other hand, for the CVD process operated on a large scale, low thermal conductivity of gas and difficulty of stirring are possible disadvantages giving a broad temperature distribution in a reactor.

From the viewpoint of equipment cost, the CVD method has both advantages and disadvantages. Usually, a large vessel applicable to pressurized or evacuated conditions is necessary for the vapor phase method. For the liquid phase method, it is an intrinsic disadvantage to deal with a large amount of solvent, which is frequently flammable, harmful, or consuming huge energy for removal; cost for treating such a solvent is also a disadvantage. However, ordinary chemical manufacturers have already had reactors useful for liquid phase reactions, and therefore it is considered that the CVD method requires relatively high cost for the plant construction in some cases.

From these reasons, the deposition of silica in a liquid phase, chemical liquid deposition (CLD), is also studied with a purpose of the enhancement of the shape selectivity. In the CLD method, not only the alkoxides $Si(OCH_3)_4$ and $Si(OC_2H_5)_4$ but also the halide $SiCl_4$, $TiCl_4$, and $SbCl_5$ are used as the regent to control the pore-opening size [12].

In this method, such a solvent as heptane is added to the zeolite sample, and a small amount of the deposited reagent is added and stirred for 5 h at the ambient temperature, followed by the calcination at 773 K. Repeated CLD is required, when the degree of the silica deposition is increased. The amount of deposited silica for the CLD modification is, however, not reported precisely in previous studies.

Thus, prepared zeolite shows the excellent shape selectivity in the adsorption of alkylaromatics. Priority of the CLD to the CVD method in the adsorption selectivity is claimed in the literature. In addition, $SiCl_4$ is reported to be more effective than $Si(OCH_3)_4$. It is interesting that the NaY is used as the starting zeolite [13], because the H-Y zeolite is difficult to be controlled by the CVD of silica. A higher reactivity of the halide compound than the alkoxide is estimated, and therefore such a halide as $SiCl_4$ could be available for the deposition on the Na-zeolite. Influence by the CLD of tetraethoxysilane on the solid acidity of HZSM-5 is studied. Two kinds of HZSM-5, which are large and small in the crystal size, are studied, and the control of the external surface acid site is measured by ammonia TPD [14]. The acid sites are decreased by 24 and 8.5% upon deposition of silica on the small and large HZSM-5, respectively. A relatively high degree of the passivation on these HZSM-5s is deduced to the decrease in the number of acid sites located in the pore mouth region.

The CLD is performed usually at a low temperature below 300 K using a solvent, hexane and cyclohexene, which is different from the CVD technique. Calcination at the high temperature such as 773 K with air is required for the CLD as well as the CVD. Therefore, not only the deposition of the reagent in the liquid phase is required to be controlled, but also the careful calcination at the high temperature is necessary. Content of water in the zeolite must be strictly controlled before the deposition, because the water present readily reacts with the CVD reagent such as $SiCl_4$ and

Si(OR)$_4$. Otherwise, the site (or position) of the deposition is not controlled, and therefore the selective modification becomes difficult. When the conditions of the deposition reaction and the modified zeolite are controlled sufficiently in the preparation of either CVD or CLD, the controlled zeolites should show similar property and selectivity, and any preference of each method will not be indicated from a scientific view. Application of the broad region of the temperature in the CVD, from room temperature up to 773 K, can be indicated as a priority of the vapor phase method.

7.2 Selective Cracking of Linear Alkane (Dewaxing)

Selective cracking of linear alkane is known as a process of dewaxing, because the linear alkane with a high freezing point is selectively removed. Industrialized processes have been developed by Mobil and Akzo [15] Co. On the other hand, CVD of silica is utilized adequately to prepare the modified mordenite with an excellent high selectivity [16]. Figure 7.7 shows an increase of the selectivity for the cracking of octane in preference to 2,2,4-trimethylpentane (isooctane), measured by a pulse reaction at 573 K. On the HM (H-mordenite) unmodified, both octane and isooctane are converted to small hydrocarbons in a similar degree. The reaction behavior is dramatically changed by the deposition of silica. On the 0.9 wt% deposition of silica, the conversion of isooctane becomes neglected almost completely, and only octane reacts. The generation of the selectivity is obviously due to the difference in molecular sizes, i.e., 0.43 and 0.62 nm for octane and isooctane, respectively. Only

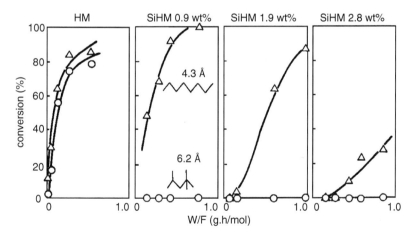

Fig. 7.7 Cracking of octane (*open triangle*) and 2,2,4-trimethylpentane (isooctane) (*open circle*) measured by a pulse method at 573 K on mordenite unmodified and modified by the deposition of silica

the smaller alkane octane reacts and the larger alkane isooctane does not react at all. Generation of such a fine reactant shape selectivity is caused by the precise control of the pore-opening size of mordenite. The size of pore opening is controlled by the deposited silica so clearly and only the small alkane is allowed to enter the pore of the modified mordenite. Furthermore, selective cracking for 3-methylheptane in preference to 2,2,4-trimethylpentane is observed also by the deposition of at 0.8 wt% silica (not shown). Thus, further precise control of the pore-opening size is attained to discriminate between the reactivities of molecules having the different molecular sizes by 0.07 nm. The generated excellent selectivity shows us a high potentiality of the CVD of silica to control the pore-opening size of zeolite.

On the modified mordenite, not only the reactant alkanes but also the product alkanes are controlled. The fraction of produced branched C_4 hydrocarbons (isobutane and isobutene) in the cracking of octane and 3-methylheptane is used as a parameter to show the distribution of products in Fig. 7.8. The fraction of the branched C_4 is as much as 50% on the HMs unmodified and modified with less than 0.8 wt% silica deposited, but it decreases to about 20% on the HM modified with more than 0.9 wt% silica deposited. By the deposition of silica, products of linear hydrocarbon increase in place of the branched hydrocarbon products. The change of product distribution is also due to the narrowing of the pore-opening size by the deposited silica. Because of the controlled pore-opening size, the formation or the diffusion of branched paraffins is strongly suppressed. These hydrocarbons diffuse out from the pore only after isomerization into smaller linear hydrocarbons. Therefore, the product shape selectivity is also realized clearly on the silica deposited mordenite.

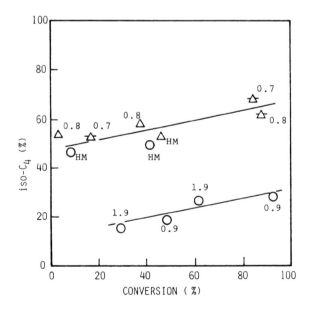

Fig. 7.8 Fraction of produced branched C_4 hydrocarbons in the cracking of octane (*open circle*) and 3-methylheptane (*open triangle*): *HM*, H-mordenite; *number*, deposited silica in wt%

7.3 Various Applications

7.3.1 Preferential Production of Dimethylamine from Methanol and Ammonia

Selective amination of methanol to dimethylamine-rich methylamines over modified mordenite is first developed by Nitto Chemical. The reaction of methanol and ammonia over an acid catalyst yields three kinds of isomers, mono-, di-, and tri-methylamine, i.e.,

$$CH_3OH + NH_3 \rightarrow NH_2(CH_3) + H_2O$$
$$CH_3OH + NH_2(CH_3) \rightarrow NH(CH_3)_2 + H_2O \quad (7.2)$$
$$CH_3OH + NH(CH_3)_2 \rightarrow N(CH_3)_3 + H_2O.$$

Trimethylamine is formed preferentially due to the thermodynamic equilibrium, but the formed composition of three isomers does not meet the demand ratio, because dimethylamine is the most desirable product. Using mordenite modified by steaming, Nitto Chemical successfully develops the process to produce dimethylamine preferentially [17]. It is expected that the steaming of mordenite controls the external surface acidity and/or the pore-opening size to some extent.

Sizes of three isomers increase in the sequence, mono- < di- < tri-methylamine. The selectivity to produce three isomers, therefore, can be controlled by adjusting the size of pore opening of zeolites. Various studies have been reported by the deposition of silica on zeolites to realize the product shape selectivity.

Bergna et al. of du Pont use Chabazite, Rho, and ZK-5 in which the pore consists of eight ring, and the size is similar to the molecular size of methylamine [18]. They modify these zeolites by the silica coating with $Si(OC_2H_5)_4$ in the liquid phase to change the products ratio. On the other hand, Segawa modifies mordenite by the CVD of silica using $SiCl_4$ vapor on Na-mordenite, followed by ion exchange to the ammonium form [19]. Kiyoura of Mitsui Chemical also uses mordenite for this reaction, and utilizes a method of CLD with $Si(OC_2H_5)_4$ for the modification. According to his review, the modification in the liquid phase CLD is preferable to CVD conducted in the vapor phase, because the vapor phase modification needs the cost for construction of the CVD reactor [20]. It is interesting for them to use mordenite with a large pore of 12-ring, because the pore size of mordenite is much larger than the molecular size of methylamine compounds, and the high degree control of the pore-opening size is required. Kiyoura discusses the catalysis on the acid site located in the eight ring of mordenite [21]. However, the reaction within the eight ring of mordenite deactivates quickly, as described in Chap. 3. Such an aspect of the silica deposition to control the pore size should be studied in more detail.

7.3.2 Improvement of the Life and the Activity of Catalysts

Pd loaded on HZSM-5 and H-mordenite is active for the selective reduction of NO with methane in the presence of oxygen. However, water co-existed retards the selective reduction activity irreversibly, and this is the serious drawback of the Pd catalyst. CVD of silica on the external surface of HZSM-5 and H-mordenite improves the activity conducted in the presence of water vapor, as shown in Fig. 7.9 [22]. The deposition of silica recovers the NO reduction activity that is deteriorated by the water vapor. The interesting effect by the deposited silica to improve the reduction activity is observed only in the NO reduction conducted in the presence of 10% water vapor. Various characterizations and knowledge of the reaction mechanism are required to understand the profile sufficiently. The finding is, however, explained due to the retardation of Pd sintering on the external surface because of the hydrophobic property created by the deposited silica. PdO interacted with the Brønsted acidity in the pore of zeolite is an active site for the NO reduction, as mentioned in Chap. 8. The Pd oxide or hydroxide migrates into the exterior of the zeolite in the presence of water to become large particles of Pd, and the catalytic activity is lost irreversibly, i.e.,

$$\text{Pd (inside)} \rightarrow \text{Pd (outside)} \rightarrow \text{Pd (sintered in } H_2O \text{ and } O_2). \tag{7.3}$$

Agglomeration into the large particle in the second step takes place readily in the presence of water and oxygen. It is shown in Fig. 7.10 that the adsorption of water is partially retarded by the silica deposited. Agglomeration of the Pd species on the external surface is the main reason of the irreversible deactivation, and the silica deposited retards the agglomeration, because the adsorption of water is retarded. This is an unusual utilization of the silica CVD for the improvement of catalyst.

A recently reported study of sulfur tolerant hydrogenation catalyst of PtNaA [51] is designed based on the similar idea, because the Pt is embedded in the pore whose

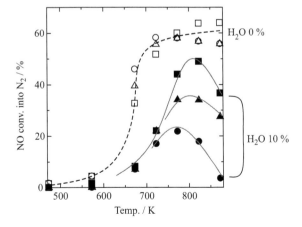

Fig. 7.9 Improvement of the Pd loaded on HZSM-5 by the deposition of 0 (*open circle, filled circle*), 4 (*open triangle, filled triangle*), and 8 (*open square, filled square*) wt% silica for the selective reduction of NO with methane in the presence of oxygen conducted without (open) and with (close) 10% water vapor

7.3 Various Applications

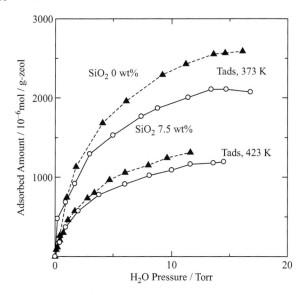

Fig. 7.10 Adsorption isotherm of water at 373 and 423 K on the Pd/HZSM-5 unmodified (*filled triangle*) and modified by 7.5 wt% SiO$_2$ deposited (*open circle*)

size is regulated by the deposited silica, and the hydrogenation takes place on the external surface by hydrogen atoms spilled over from the Pt inside the pore. The Pt metal is not deactivated in the presence of H$_2$S impurity, because H$_2$S cannot enter the pore. Thus, the catalyst system consisted of interior and exterior of the A zeolite that is well designed to have the hydrogenation activity sustainable to the sulfur compound.

A heat-resisting metal oxide such as silica coated alumina is developed as a useful application of the CVD of silica to metal oxide. The surface area of alumina drops usually to less than 5 m^2 g^{-1} upon the calcination at 1,493 K, whereas the surface area is kept at 50–60 m^2 g^{-1} after coating the surface with a mono to double layer of silica [23]. Alumina is not agglomerated into the large particle even at the high temperature calcination only by coating the surface by a monolayer of silica. Thus, the deposited silica retards the agglomeration of alumina that occurs between the alumina particles. The composition and structure of the external surface seem to determine the condensation reaction of metal oxides to become the large particles.

After calcination at 1,493 K, alumina shows usually the α-phase with a small surface area. However, the silica-coated alumina shows the γ, θ, or η-phase and the large surface area. Therefore, it is found that alumina is not always stabilized as the α-phase, when it is calcined at 1,493 K. Agglomeration of the alumina particles occurs from the surface as shown by Scheme 7.1, and the transformation of the crystal phase seems to be stimulated by the surface reaction. In other words, the transformation of crystal phase does not seem to occur without agglomeration into large particles.

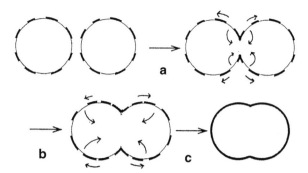

Scheme 7.1 Retard of the agglomeration of the alumina particles by the deposited silica on alumina. Sintering may occur with (**a**) formation of a grain boundary, (**b**) neck growth, and (**c**) formation of a small particle with silica coating

7.3.3 Selective Removal of Undesired Products

Selective removal of hydrogen by a combustion is studied using a mixture of H_2/isobutane on the SiCuHZSM-5 [24]. To produce methyl tertiary butyl ether (MTBE), isobutene is produced from isobutane by the dehydrogenation, and the isobutane is produced as a result of skeletal isomerization of butane. However, the conversion level of isobutane to isobutene is limited by the equilibrium. Therefore, hydrogen is selectively removed by the oxidation in a mixture of H_2/isobutene. This is a kind of oxidative dehydrogenation of isobutane to break a limit of the equilibrium in order to enhance the yield of isobutene. Thus, the idea to enhance the selectivity is to control the pore-opening size of HZSM-5 so as to allow only H_2 to enter the pore. On the SiCuHZSM-5 with 0.58 wt% silica deposited, 75% of hydrogen is removed by the oxidation, while isobutene is removed only by 1.6%. Thus, hydrogen is removed with a high selectivity.

Selective hydrogenation of acetylene is studied using a mixture of acetylene/butadiene in ethylene on the SiNiA [25]. Small amount of acetylene is contained in ethylene produced by the steam cracking of natural gas, and it may induce the explosion in the downstream cryogenic separation process. Butadiene is also contained in an impurity level, and it should remain unconverted. Therefore, the selective hydrogenation of acetylene into ethylene is studied. Finally, Ni-KA modified with $Si(OEt)_4$ is developed to have the high selectivity in the hydrogenation of acetylene.

7.3.4 Applications to Zeolites from Various View-Points

In this section, various applications of the deposition of silica are summarized. Not only the typical shape selective reactions but also various interesting reactions and

7.3 Various Applications   145

separations have been studied. The deposition of silica on the external surface not only to control the pore-opening size but also to inactivate the external surface is the principle brought by the method. Typical examples of the silica deposition, listed sequentially in the published year, are summarized in Table 7.2.

Table 7.2 Study on CVD (V) and CLD (L) of silica compounds on zeolites in 1978–2006

Zeolites	Reagents	L or V	Subjects in the study	Reference
Mordenite	SiH$_4$	V	Adsorption property	[26]
Mordenite	Si(OMe)$_4$	V	Selective cracking	[16]
H-Y	Silane derivatives	V	(NMR study on anchoring)	[27]
ZSM-5	Si(OEt)$_4$	V	p-Alkylbenzene formation	[6]
Rho, ZK-5	Si(OEt)$_4$	L	Dimethylamine formation	[18]
Mordenite, X, A	Si(OEt)$_4$	V	Gas adsorption	[28]
H-Y	Si(OEt)$_4$	V	Selective adsorption	[29]
Mordenite	Si$_2$H$_6$	V	Adsorption property	[30]
Offretite	Si(OMe)$_4$, octamethylcyclotetrasiloxane	V	Inactivation of external surface	[31]
HZSM-5	Si(OEt)$_4$	V, L	Inactivation of external surface	[32]
Hβ	Si(OEt)$_4$ or chloromethylsilane	L	Inactivation of external surface	[33]
HZSM-5	Si(OEt)$_4$, Ge(OEt)$_4$	V	Cracking, MTG	[34]
Silicalite membrane	Si(OEt)$_4$, Si(OMe)$_4$ with O$_3$	V	Separation of gases	[35]
Silica membrane	Si(OEt)$_4$	V	Preparation of membrane	[36]
H-Y	SiH$_4$ with N$_2$O	V	Disproportionation of ethylbenzene	[37, 52]
MCM-22-Mg	Si(OEt)$_4$	V	Isomerization of 1-butene into isobutene	[38]
Pt-KL	Si(OEt)$_4$, disilazane	V	Benzene formation	[39]
HZSM-5	Si(OEt)$_4$	V, L	Preparation conditions	[40]
USY	Si(OMe)$_4$	V	Control of the pore opening	[41]
Mordenite, β, ZSM-5	Si(OEt)$_4$	V, L	Mechanism	[42]
HZSM-5	Si(OEt)$_4$ in toluene and methanol	V	p-ethylphenol formation	[43]
Ga-MFI	Si(OEt)$_4$ in toluene and methanol	V	p-DEB formation	

(continued)

Table 7.2 (continued)

Zeolites	Reagents	L or V	Subjects in the study	Reference
MFI-type membrane	Me$_2$(OMe)$_2$Si, Et$_2$(OEt)$_2$Si	V	H$_2$ permeability	[44]
Mordenite	Si(OEt)$_4$	L	Disproportionation of cumene	[45]
HZSM-5	SiCl$_4$	L	Alkylation of toluene	
HZSM-5	Si(OEt)$_4$	L	Site and mechanism	[14]
MCM-41	Si(OEt)$_4$, Si(OMe)$_4$	V	Pore size control	[46]
HZSM-5	Si(OEt)$_4$ in toluene and methanol	L	Alkylation of EB	[47]
Mo/HZSM-5	Si(OEt)$_4$	V	Benzene formation	[48]
MCM-22	3-Glycidoxypropyltrimethoxysilane	L	p-Xylene formation	[49]
Mo/HZSM-5	Various Si compounds (3-aminopropyl-triethoxysilane)	L	Benzene formation	[50]
PtNaA	Si(OEt)$_4$	V	Sulfur tolerant hydrogenation	[51]

This method is applied to almost all the zeolite species. Vapor-phase deposition (CVD) is utilized often, but the liquid-phase deposition (CLD) is also utilized in a similar degree. Si(OC$_2$H$_5$)$_4$ is the deposited reagent most frequently utilized. Applications to mesoporous materials MCM-41 [46] and membrane [35] are studied with the similar purpose.

References

1. J.H. Kim, T. Kunieda, M. Niwa, J. Catal. **173**, 433 (1998)
2. M. Niwa, M. Kato, T. Hattori, Y. Murakami, J. Phys. Chem. **90**, 6233 (1986)
3. T. Hibino, M. Niwa, Y. Murakami, J. Catal. **128**, 551 (1991)
4. K. Tominaga, S. Maruoka, M. Gotoh, N. Katada, M. Niwa, Microporous Mesoporous Mater. **117**, 523 (2009)
5. J. Das, Y.S. Bhat, A.B. Halgeri, Ind. Eng. Chem. Res. **32**, 2525 (1993)
6. I. Wang, C.-L. Ay, B.-J. Lee, M.-H. Chen, Appl. Catal. **54**, 257 (1989)
7. J.H. Kim, A. Ishida, M. Okajima, M. Niwa, J. Catal. **161**, 387 (1996)
8. H.P. Roger, M. Kramer, K.P. Moller, C.T. O'Connor, Microporous Mesoporous Mater. **21**, 607 (1998)
9. I. Wang, C.-L. Ay, B.-J. Lee, M.-H. Chen, *Proceedings of 9th International Congress on Catalysis*, ed. by M.J. Phillips, M. Ternan, vol 1, p. 324, (1988)
10. A.B. Halgeri, J. Das, Catal. Today **73**, 65 (2002)
11. N. Katada, Y. Kageyama, M. Niwa, J. Phys. Chem. B **104**, 7561 (2000)
12. Y.H. Yue, Y. Tang, Y. Liu, Z. Gao, Ind. Eng. Chem. Res. **35**, 430 (1996)
13. Y.H. Yue, Y. Tang, Y. Liu, Z. Gao, in *Progress in Zeolite and Microporous Materials*, ed. by H. Chon, S.-K. Ihm, Y.S. Uh. Studies in Surface Science and Catalysis, vol 105 (Elsevier, Amsterdam, 1997) p. 2059
14. S. Zheng, H.R. Heydenrych, A. Jentys, J.A. Lercher, J. Phys. Chem. B. **106**, 9552 (2002)
15. H.W.H. Free, T. Schockaert, J.W.M. Sonnemans, Fuel Proc. Technol. 35, 111 (1993)

References

16. M. Niwa, S. Morimoto, M. Kato, T. Hattori, Y. Murakami, *Proceedings of 9th International Congress on Catalysis*, vol IV, p. 701 (1984)
17. Y. Ashina, T. Fujita, M. Fukatsu, K. Niwa, J. Yagi, *Proceedings of the 7th International Zeolite Conference* (Kodansha/Elsevier, Tokyo/Amsterdam, 1986), p. 779
18. H.E. Bergna, M. Keane Jr., D.H. Ralston, G.C. Sonnichsen, L. Abrams, R.D. Shanon,. J. Catal. **115**, 148 (1989)
19. K. Segawa, H. Tachibana, J. Catal. **131**, 482 (1991)
20. T. Kiyoura, Shokubai **40**, 287 (1988)
21. T. Kiyoura, J. Catal. **170**, 204 (1997)
22. M. Suzuki, J. Amano, M. Niwa, Microporous Mesoporous Mater. **21**, 541 (1998)
23. N. Katada, H. Ishiguro, K. Muro, M. Niwa, Chem. Vap. Dep. **1**, 54 (1995)
24. C.H. Lin, K.C. Lee, B.Z. Wan, Appl. Catal. A Gen. **164**, 59 (1997)
25. D.R. Corbin, L. Abrams, C. Bonifaz, J. Catal. **115**, 420 (1989)
26. R.M. Barrer, E.F. Vansant, G. Peeters, J. Chem. Soc., Faraday I **74**, 1871 (1978)
27. T. Bein, R.F. Carver, R.D. Farlee, G.D. Stucky, J. Am. Chem. Soc. **110**, 4546 (1988)
28. Y. Teraoka, K. Kunitake, S. Kagawa, M. Iwamoto, Nippon Kagaku Kaishi 424 (1989)
29. H. Itoh, S. Okamoto, A. Furuta, Nippon Kagaku Kaishi 420 (1989)
30. Y. Yan, E.F. Vansant, J. Phys. Chem. **94**, 2582 (1990)
31. M. Chamoumi, D. Brunel, F. Fajula, P. Geneste, P. Moreau, J. Solofo, Zeolites **14**, 282 (1994)
32. R.W. Weber, J.C.Q. Fletcher, K.P. Moller, C.T. O'Connor, Microporous Mater. **7**, 15 (1996)
33. P.J. Kunkeler, D. Moeskops, H. van Bekkum, Microporous Mater. **11**, 313 (1997)
34. P. Tynjala, T.T. Pakkanen, J. Mol. Catal. A Chem. **122**, 159 (1997)
35. M. Nomura, T. Yamaguchi, S. Nakao, Ind. Eng. Chem. Res. **36**, 4217 (1997)
36. B.-K. Sea, K. Kusakabe, S. Morooka, J. Memb. Sci. **130**, 41 (1997)
37. E. Klemm, M. Seitz, H. Scheidat, G. Emig, J. Catal. **173**, 177 (1998)
38. S.H. Baeck, K.M. Lee, W.Y. Lee, Catal. Lett. **52**, 221 (1998)
39. J. Zheng, Y. Chun, J. Dong, Q. Xu, J. Mol. Catal. A Chem. **130**, 271 (1998)
40. H.P. Roger, M. Kramer, K.P. Moller, C.T. O'Connor, Microporous Mesoporous Mater. **23**, 179 (1998)
41. J.H. Kim, Y. Ikoma, M. Niwa, Microporous Mesoporous Mater. **32**, 37 (1999)
42. R.W. Weber, K.P. Moller, C.T. O'Connor, Microporous Mesoporous Mater. **35**, 533 (2000)
43. J. Das, A.B. Halgeri, Appl. Catal. A Gen. **194**, 359 (2000)
44. T. Masuda, N. Fukumoto, M. Kitamura, S.R. Mukai, K. Hashimoto et al. Microporous Mesoporous Mater. **48**, 239 (2001)
45. T.-W. Kuo, C.-S. Tan, Ind. Eng. Chem. Res. **40**, 4724 (2001)
46. K. Fodor, J.H. Bitter, K.P. de Jong, Microporous Mesoporous Mater. **56**, 101 (2002)
47. N. Sharanappa, S. Pai, V.V. Bokade, J. Mol. Catal. A Chem. **217**, 185 (2004)
48. H. Liu, Y. Li, W. Shen, X. Bar, Y. Xu, Catal. Today **93**, 65 (2004)
49. X. Ren, J. Liang, J. Wang, J. Porous Mater. **13**, 353 (2006)
50. S. Kikuchi, R. Kojima, H. Ma, J. Bai, M. Ichikawa, J. Catal. **242**, 349 (2006)
51. H. Yang, H. Chen, J. Chen, O. Omotoso, Z. Ring, J. Catal. **243**, 36 (2006)
52. M. Seitz, E. Klemm, G. Emig, *Proceedings of the 7th International Zeolite Conference*, MRS (1999), p. 1969

Chapter 8
Zeolite Loading Property for Active Sites and XAFS Measurements

Abstract New methodologies for X-ray absorption fine structure (XAFS) measurements, such as Quick XAFS and Dispersive XAFS, are reviewed. These techniques are applied to observe the dynamic structural change of Pd, i.e., clustering and dispersion, which are caused by the strong interaction with acid sites of zeolites of ZSM-5, mordenite, H-Y, and USY.

8.1 EXAFS and XANES Measurements of Loaded Metals

8.1.1 DXAFS and QXAFS Analysis

X-ray absorption fine structure (XAFS) is a useful technique in the analysis of local structure of heterogeneous catalysts, whose structural information is otherwise difficult to be obtained. Using XAFS technique, valuable information such as local structure, symmetry, and valence sate around specific elements can be obtained. In the conventional XAFS technique, data collection has been carried out under the static conditions using double monochromators that is moved stepwise. Therefore, long time period up to 1 h is required to collect XAFS data. Recent development in XAFS technique including Quick-XAFS (QXAFS) and Energy-Dispersive XAFS (DXAFS) realized the high-speed measurement of XAFS data. Unlike the conventional method, the double monochromator is moved continuously to obtain monochromatic X-ray beam in the QXAFS, which enables us to obtain data in several seconds–minutes. On the other hand, in the DXAFS, intensities in whole energy region are collected at the same time using a bent polychromator as displayed in Fig. 8.1. Bragg and Laue configurations are chosen depending on the X-ray energy region at $E_{X\text{-ray}} < 12$ keV and at $E_{X\text{-ray}} > 12$ keV for XAFS measurements at Japan Synchrotron Radiation Research Institute (SPring-8), respectively. At higher energy region, Bragg configuration is preferred because X-ray goes into the deeper position of the polychromator so that the energy resolution of spectra becomes lower. Using these techniques combined with an appropriate cell and a gas-flow line, time-resolved in situ measurements of the various chemical processes, such as the formation process of

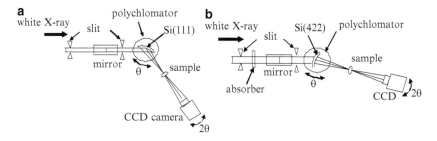

Fig. 8.1 An energy-dispersive XAFS instrument installed in SPring-8 BL28B2. Bragg ($E_{\text{X-ray}} < 12\,\text{keV}$) (**a**) and Laue configurations ($E_{\text{X-ray}} > 12\,\text{keV}$) (**b**)

the active sites in heterogeneous catalysts and clustering of metal atoms, are realized. In recent years, much effort has been devoted to design various in situ cells and studies combined with other techniques, i.e., XRD and a mass spectrometer. In this chapter, we will focus on the studies concerning the structural change of Pd induced by the metal–support interaction with zeolite supports. Small metal clusters occluded in a zeolite pore have been studied primarily from the viewpoint of the formation of uniform active sites for catalytic reactions. The introduction of metals in zeolites has been achieved through various techniques including chemical vapor deposition, ship-in-bottle method and decarbonylation of carbonyl clusters in zeolite pores [1–5]. In particular, Pd clusters in zeolites have been paid attention since Pd exhibits high catalytic activities in many kinds of valuable reactions, such as selective reduction of NO, total combustion of hydrocarbons, and organic reactions. Influence by not only Brønsted acid site but also the structure of zeolite on the generation of metal Pd clusters and highly dispersed PdO is studied. This is because it has been revealed that the structure and the acid sites of zeolites considerably affect the generation of active sites for catalytic reactions. As a matter of fact, the catalytic performance of Pd greatly depends on the structure and composition of zeolite supports. The fact suggests that the strong metal–support interaction between PdO and Brønsted acid sites plays an important role in not only the generation but also the catalytic performance of active Pd center. The interaction is directly evidenced from the structural change of Pd induced by Brønsted acid site of zeolites in the oxidative or reductive atmosphere. However, the formation and structure of the active Pd species or its precursor and the role of the Brønsted acid sites associated with Pd are rather ambiguous. Here, in order to precisely reveal the metal-support interaction, DXAFS and QXAFS techniques are utilized to follow the dynamic structural change of Pd induced by the interaction with Brønsted acid sites of zeolites.

8.1.2 Formation of Molecular-Like PdO Through the Interaction with Acid Sites of Zeolites

It has been recognized that the catalytic activity of Pd changes significantly depending on the employed supports. For instance, Pd loaded on various supports that

8.1 EXAFS and XANES Measurements of Loaded Metals

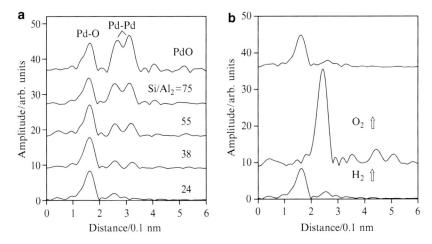

Fig. 8.2 k^3-weighted Pd K-edge EXAFS Fourier transforms of 0.4 wt% Pd loaded on H-ZSM-5 with different Al concentrations (**a**) and H-ZSM-5 (Si/Al$_2$ = 24) treated with H$_2$, and O$_2$ at 773 K (**b**)

have acid character exhibits high activities in the selective reduction of NO with methane in the presence of excessive O$_2$. In contrast, the total oxidation of methane takes place over the Pd loaded on the supports which lacks acid character. In order to reveal the origin in the effect by acidity of support, H-ZSM-5 zeolites with different Al content are employed as supports for Pd and the structure of Pd is analyzed by means of Pd K-edge EXAFS. Figure 8.2a shows the Fourier transforms of the $k^3\chi(k)$ EXAFS for Pd/H-ZSM-5 with different Al content. All samples are oxidized under an oxygen flow at 773 K for 3 h. The peaks appeared at 0.26 and 0.31 nm are assignable to the Pd–Pd bonds characteristic of agglomerated PdO, as can be seen from the comparison with the spectrum of PdO. The intensity of these Pd–Pd peaks decreases accompanied by an increase in the Al content of H-ZSM-5 (decrease in the Si/Al$_2$ ratio). The Pd–Pd bond completely disappears on the Pd/H-ZSM-5 with the highest Al content (Si/Al$_2$ = 23.8), and the catalyst exhibits a high activity in the selective reduction of NO with methane. The results indicate that the size of PdO is a function of the amount of Brønsted acid in H-ZSM-5 and it decreases with increase in the acid amount of H-ZSM-5, since the intensity of Pd–Pd shell reflects the size of PdO. On the other hand, as for Pd–O bond observed at 0.16 nm in Fig. 8.2a, the spectra for bulk PdO and highly dispersed PdO on H-ZSM-5 are quite similar. In addition, the coordination number and bond distance of Pd–O determined by the curve fitting analysis on highly dispersed PdO agree well with those on bulk PdO, implying that the local structure of highly dispersed PdO is closely similar to that of bulk PdO. Therefore, it can be noted that the role of Brønsted acid sites of H-ZSM-5 is not to provide the ion-exchange sites for Pd^{2+}, but to stabilize the dispersed state of PdO. Based on the EXAFS analysis, the local structure of Pd in the oxidized Pd/H-ZSM-5 is proposed by Wang and Liu, where Pd is surrounded by four oxygen atoms in a square planar, a part of which comes from the zeolite structure as displayed in Fig. 8.3 [6].

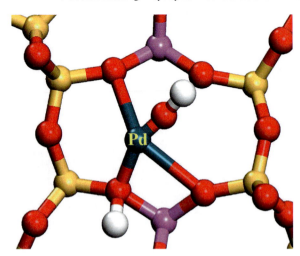

Fig. 8.3 Schematic structure of PdO interacted with Brønsted acid sites of H-beta proposed by Liu et al.

In order to confirm the ability for Brønsted acid sites of zeolite to anchor PdO, the regeneration of dispersed PdO upon the repetition of reduction and oxidation treatments are followed by EXAFS. The experiment is conducted on the Pd/H-ZSM-5 ($Si/Al_2 = 23.8$) where a highly dispersed PdO is observed by the oxidation treatment as explained above. Figure 8.2b shows the EXAFS FT spectrum measured after the reduction of previously oxidized Pd/H-ZSM-5. The formation of metal Pd is confirmed from the appearance of an intense peak at 0.24 nm. The particle size of the metal Pd calculated from the Pd–Pd coordination number (CN = 10.6) is estimated to be >3 nm, which is far larger than the zeolite pore diameter. The change in the spectra indicates that the highly dispersed PdO is reduced and migrated to form aggregated Pd metal particle on the external surface of zeolite. The reduced sample is subsequently oxidized under the oxygen flow at 773 K for 3 h again. The spectrum measured after oxidation is completely identical to that measured after initial oxidation treatment as included in Fig. 8.2b. Therefore, it can be mentioned that the aggregated Pd returns to the highly dispersed PdO in H-ZSM-5 pore. This behavior of Pd demonstrates the high mobility of PdO and the presence of a strong interaction between Brønsted acid sites of H-ZSM-5 and PdO. Probably, the acid-base interaction between the highly dispersed PdO and Brønsted acid sites of zeolite promotes the disruption and fixation of highly dispersed PdO.

8.1.3 Reversible Cluster Formation Through the Interaction with Acid Sites of Zeolites

In order to further reveal the dynamic behavior of Pd with zeolite supports, the agglomeration process of Pd^0 is measured in the atmosphere of H_2. For these pur-

8.1 EXAFS and XANES Measurements of Loaded Metals 153

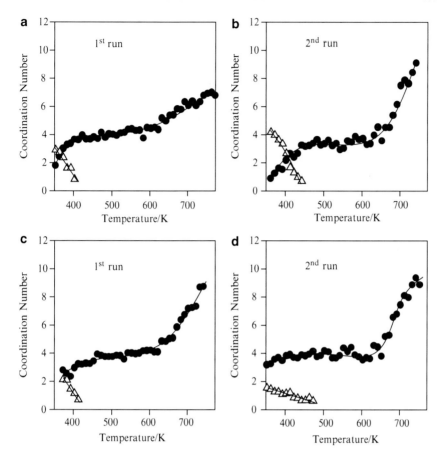

Fig. 8.4 Dependence of coordination number of Pd–Pd (*filled circle*) and Pd–O (*open triangle*) on the temperature measured in an 8% H_2 flow; Pd/H-ZSM-5 (**a, b**); Pd/H-Mordenite (**c, d**)

poses, in situ DXAFS experiment is applied to determine the local structure of Pd during the temperature-programmed reduction in the atmosphere of H_2. Figure 8.4a, c shows the coordination number (CN) of nearest neighboring Pd–Pd (metal Pd) peak calculated based on the curve-fitting analysis of the EXAFS spectra for H-ZSM-5 (Si/Al$_2$ = 23.8) and H-Mordenite (Si/Al$_2$ = 20), respectively. A slight increase in the Pd–Pd is observed from the beginning of the reduction. At the same time, the CN of Pd–O bond decreases, suggesting that the reduction of PdO to metal Pd takes place up to 440 K. After the completion of the reduction, the CN of metal Pd–Pd keeps a constant value of 4 between 430 and 620 K on both H-ZSM-5 and H-Mordenite. It is obviously noted that the appearance of the plateau means the generation of a stable Pd cluster at the temperature region. From the CN value, the Pd cluster is estimated to consist of 6 atoms. On further heating the samples under flowing H_2, the CN steeply increases from 623 to 773 K. Probably, the Pd$_6$ cluster migrates into the external surface of zeolites to form the agglomerated metal Pd

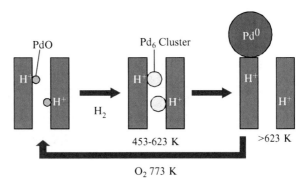

Fig. 8.5 Reversible structural change of Pd induced by the interaction with acid sites of ZSM-5 and Mordenite zeolites

particles. The change in CN on Pd/H-Mordenite is similar to that on Pd/H-ZSM-5, except that the degree of growth of metal Pd observed above 623 K is steeper on H-Mordenite. After the measurements given in Fig. 8.4a, c, the samples are oxidized at 773 K for 4 h in an O_2 flow. Then the in situ cell is cooling down to the room temperature and the temperature-programmed reduction is carried out in an 8% H_2 flow again. Figure 8.4b, d shows the CN of Pd on H-ZSM-5 and H-Mordenite measured during the second run, respectively. A similar pattern in the change of CN (Pd–Pd) to the first runs is observed on both H-ZSM-5 and H-Mordenite. That is to say, a plateau of CN (Pd–Pd) is observed from 453 to 623 K similarly to the first run of the experiment. Therefore, it is confirmed that the generation of a stable metal Pd cluster is reversible upon the oxidation with O_2 at 773 K and successive reduction with H_2, as illustrated in Fig. 8.5. The phenomena could be understood when the fact is taken into consideration that the agglomerated metal Pd is readily re-dispersed onto acid sites of zeolites through the oxidation at 773 K in an O_2 flow evidenced in the previous section.

8.2 In Situ QXAFS Studies on the Dynamic Coalescence and Dispersion Processes of Pd in USY Zeolite

Pd or bimetallic Pd–Pt supported on the USY zeolite has been found to exhibit high sulfur tolerance in the hydrogenation of aromatics and hydrodesulfurization [7]. Moreover, Pd clusters and atomic Pd generated in the supercage of an FAU-type zeolite are active and reusable in Heck and Suzuki-Miyaura reactions as will be mentioned in Chap. 9 [8]. In order to reveal the genesis of active Pd species and dynamic behaviors of Pd in the pore of a USY zeolite, QXAFS technique is first applied to detect the detailed structural change of Pd in the USY zeolite. In general, QXAFS is suitable to monitor a relatively slow structural change that occurs in seconds to minutes, while DXAFS is applicable to measure structural change in less

8.2 QXAFS Studies on the Dynamic Coalescence and Dispersion Processes

than seconds. In addition to this, data with high-energy resolution are possible to obtain by QXAFS because usual double monochromator is utilized in the QXAFS mode.

Fourier transforms of the $k^3\chi(k)$ data of 0.4 wt%–Pd(NH$_3$)$_4$Cl$_2$/H-USY collected after the admission of H$_2$ at room temperature are given in Fig. 8.6. The intensity of the Pd–N peak that appears at 0.16 nm at the initial stage gradually decreases with time. This is accompanied by an increase in a new peak attributable to the Pd–Pd bond of Pd metal at 0.24 nm as a result of the reduction of Pd^{2+} to give metal Pd clusters. Then the flowing gas is switched to 8% O$_2$ for 20 min, followed again by 8% H$_2$; at this point, the second QXAFS measurement is carried out. In the second step, a small Pd–O bond could be seen at the initial stage, indicating that the metal Pd clusters generated in the first step are partially oxidized by the exposure to 8% O$_2$. The intensity of the Pd–O bond decreases while that of the Pd–Pd bond increases quickly within 3 min after switching to 8% H$_2$. It is clear that the intensity of the metal Pd–Pd bond becomes larger than that in the first step. A similar change is observed on further switching the flowing gas to O$_2$ and then to H$_2$ (third and fourth steps). A comparison of the spectra at 20 min reveals that the intensity of the Pd–Pd bond increases in a stepwise fashion, meaning that the sizes of the Pd clusters increases with the repetition of the O$_2$ and H$_2$ exposures as illustrated in Fig. 8.7. In other words, it is possible to regulate the size of Pd clusters simply by changing the H$_2$-exposure times for Pd/USY.

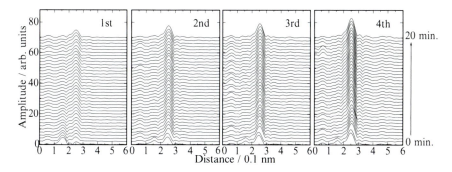

Fig. 8.6 Pd K-edge EXAFS Fourier transforms for Pd/H-USY measured in the atmosphere of 8% H$_2$. Fourier transforms range, 30–130 nm^{-1}

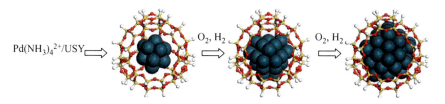

Fig. 8.7 A proposed stepwise growth of Pd clusters in H-USY at room temperature

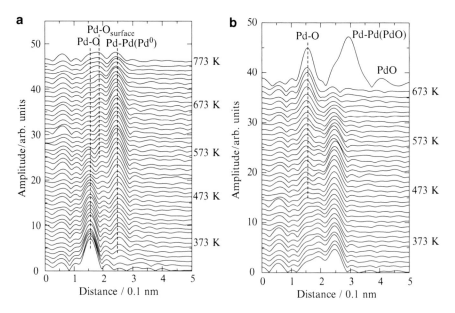

Fig. 8.8 Pd K-edge EXAFS Fourier transforms for Pd/USY measured in the TPR (**a**) and TPO (**b**)

Then TPR and TPO are repeated one after another to follow the structural change of Pd induced by reduction and oxidation at elevated temperature; first, the changes in the structure of Pd loaded on the USY zeolite is measured by QXAFS in an atmosphere of hydrogen (TPR). The Fourier transforms of the $k^3\chi(k)$ EXAFS collected after every 10 K are given in Fig. 8.8a. In the beginning, the Pd–O appeared at 0.16 nm gradually decreases. This is accompanied by an increase in the temperature, and the new peak attributable to the Pd–Pd bond of Pd metal appears at 0.24 nm as a result of the reduction of the dispersed PdO. On further increasing the temperature above 673 K a new peak appears at 0.18 nm, which is assignable to the oxygen in the framework of the USY zeolite (denoted as Pd–O$_{surface}$) by considering that Pd is already reduced to Pd^0. The CNs of these bonds are determined by a curve-fitting analysis and thus obtained data are summarized in Fig. 8.9a. In the initial step of the first TPR, the CN of the Pd–O bond decreases up to 523 K, while the CN of Pd–Pd bond increased up to 7.5 at 673 K. The CN of the Pd–Pd bond decreases on a further increase in the temperature, indicating the dispersion of previously agglomerated Pd metal at elevated temperatures. At the same time, the CN of Pd–O$_{surface}$ bond increases by an increase in the temperature. This fact means that a strong interaction between the framework of the USY zeolite and Pd leads to the dispersion of the Pd metal. This phenomenon appears to be interesting, taking into account that the heating at a high temperature usually results in the severe sintering of the metal. The sample is cooled down to room temperature and the TPO experiment is subsequently carried out after switching the flowing gas with an 8% O$_2$ flow. The Fourier transforms of the $k^3\chi(k)$ EXAFS collected after every 10 K are given in Fig. 8.8b and the CNs determined from these spectra are given in Fig. 8.9b. The

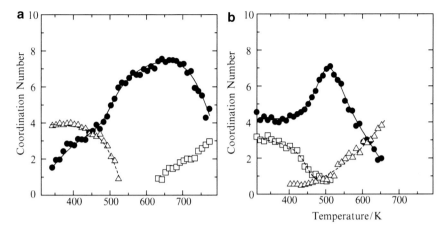

Fig. 8.9 Coordination numbers determined by curve-fitting analysis plotted as a function of temperature measured in TPR (**a**) and TPO (**b**) (*filled circle*) Pd–Pd (metal), *open triangle* Pd–O, *open square* Pd–O$_{surface}$

CN of Pd–O$_{surface}$ decreases in the initial stage, while the CN of the Pd–Pd reached to 7.1 at 513 K. This change implies that the removal of the Pd metal clusters from the framework of the zeolite and the agglomeration takes place in the initial step. On further increasing the temperature (>513 K), the CNs of Pd–Pd decreases and in turn, the CN of covalent Pd–O increases. Even though the disappearance of the Pd–Pd bond attributable to the Pd metal at 773 K, the Pd–Pd bond of PdO does not appear, suggesting the formation of highly dispersed PdO. Based on these QXAFS analysis of the Pd/USY measured during TPR–TPO cycles, the structural change of Pd occurred in the USY zeolite at elevated temperature is proposed in Fig. 8.10. In the TPR process, dispersed PdO is reduced to Pd^0, followed by agglomeration to give Pd clusters inside the supercage of USY; the formed cluster is further dispersed into sodalite cages. In the subsequent TPO, the Pd^0 clusters migrate to be dispersed onto the acid sites as the dispersed PdO form.

8.3 Formation of the Atomically Dispersed Pd^0 Through H$_2$ Bubbling in *o*-Xylene: XAFS Measurements of Metals in the Liquid [9]

Although in situ XAFS measurements of catalysts have primarily been carried out in a gas phase, measurements in liquid phase seems to be important to obtain insight into the active species generated in a solvent, taking into account that many of the catalytic reactions are performed in the liquid phase. However, little is known about the genesis and structure of metal species in the liquid phase. This is probably because the direct characterization of a solid catalyst present in a solvent is difficult in general. At this point, an XAFS technique is suitable for the characterization of

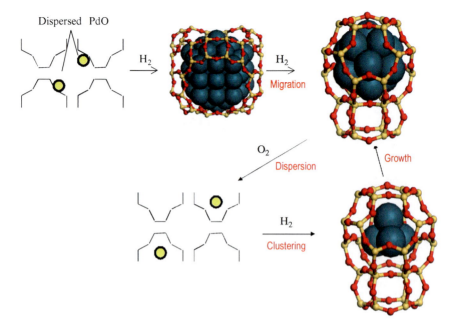

Fig. 8.10 A proposed structural change of Pd in the course of temperature-programmed reduction and oxidation

the active metal species generated even in solvents. The XAFS technique employing synchrotron radiation is particularly suitable for the in situ observation of supported Pd catalysts in the liquid phase, even with such a low loading as 0.4 wt%. In addition, the fact that X-ray absorption by zeolites and solvents is small in comparison with that of Pd at the energy of Pd K-edge (24.3 keV) is beneficial for collecting high-quality data. Here, the Pd species obtained by H_2 bubbling in o-xylene are analyzed by in situ XAFS. Pd K-edge EXAFS data collection is carried out under in situ conditions using the cell shown in Fig. 8.11. This in situ cell is made of PET resin to enable the penetration of X-ray. The light pass length (thickness of the cell) is chosen to be 30 mm. H_2 diluted with Ar is flown into the catalyst immersed in a solvent, while the solution is heated and stirred vigorously with a hot stirrer that is placed under the in situ cell.

Figure 8.12 shows the Pd K-edge $k^3 \chi(k)$ EXAFS and their Fourier transforms of $Pd(NH_3)_4Cl_2$/USY measured with 6%-H_2 bubbling in o-xylene at different temperatures. The spectra measured after H_2 bubbling at 323–353 K (b, c) are close to that of the as-received one (a), in which the Pd–N bond is observed at 0.15 nm. This implies that the Pd remains intact at these temperatures. The spectra changes significantly after H_2 bubbling at 373–383 K; two small peaks appears at 0.17 and 0.22 nm (d, e, phase shift uncorrected). At the same time, the color of the sample changes from white to dark brown. Results of the curve-fitting analysis reveals that these peaks are assignable to Pd–$O_{zeolite}$ and Pd–Al(Si) due to the framework of the USY-zeolite, respectively, where the length of the Pd–$O_{zeolite}$ bond (0.216 nm) is

8.3 Formation of the Atomically Dispersed Pd⁰ Through H$_2$ Bubbling in o-Xylene 159

Fig. 8.11 The equipments for in situ XAFS measurements in o-xylene

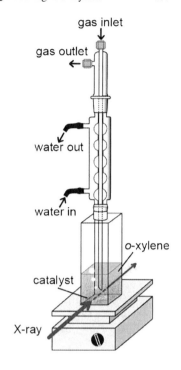

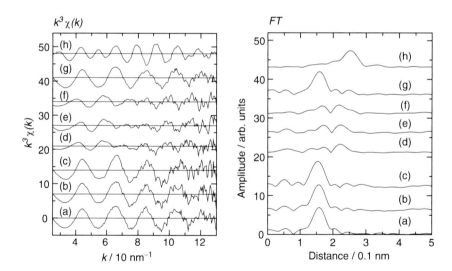

Fig. 8.12 Pd K-edge EXAFS $k^3\chi(k)$ (*left*) and their Fourier transforms (*right*) of Pd(NH$_3$)$_4$Cl$_2$/USY. Initially (**a**); after bubbling with 6% H$_2$ at 323 K (**b**), 353 K (**c**), 373 K (**d**), 383 K (**e**); air was exposed to the sample (**e**) at r.t. (**f**); air was exposed to the sample (**e**) after the removal of o-xylene at r.t. (**g**); 6% H$_2$ was exposed to the 0.4 wt%-Pd(NH$_3$)$_4$Cl$_2$/USY at r.t. without solvent (**h**)

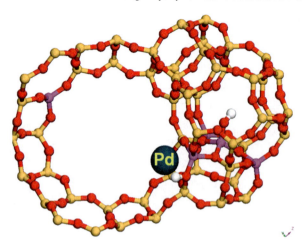

Fig. 8.13 Proposed structure of monomeric Pd generated in the supercage of USY zeolite

longer than that of the covalent Pd–O bond (0.202 nm) in PdO. The absence of the Pd–Pd bond indicates the formation of monomeric Pd. Based on the XAFS analysis, a possible structure of monomeric-Pd/USY is proposed, as shown in Fig. 8.13, in which the Pd atom is stabilized on the acid site (site III). The EXAFS of Pd/USY exposed to 6% H_2 at room temperature (Pd/USY is reduced with 6% H_2 without solvent) is given in Fig. 8.12h. The features are significantly different from those measured in the presence of o-xylene. That is to say, Pd^{2+} is readily reduced with 6% H_2 to give Pd clusters with CN(Pd–Pd) = 6.1, which corresponds to ca. 13 atoms. This marked difference in the reducing manner of Pd^{2+} implies that o-xylene retards the reduction of Pd^{2+}, which leads to the formation of a monomeric Pd. In order to examine the stability of the monomeric Pd generated in o-xylene, the monomeric-Pd/USY is exposed to air at room temperature, while it is immersed in o-xylene. The spectrum measured after exposure to air for 2 weeks (f) is identical to that of monomeric Pd (e), as shown in Fig. 8.12. However, after the removal of o-xylene by filtration and drying in air, the monomeric Pd is quickly oxidized to yield dispersed PdO, which is confirmed from the appearance of the single Pd–O bond (g). Therefore, the monomeric Pd could stably exist only when it is immersed in o-xylene.

Figure 8.14 shows the Pd K-edge EXAFS of $Pd(NH_3)_4Cl_2$/USY is measured in o-xylene after H_2 bubbling with different partial pressures of H_2. For this experiment, H_2 diluted with Ar is introduced to the in situ cell, while the total flow rate is kept at 30 ml min^{-1}. The reduction of Pd^{2+} was incomplete after the bubbling with Ar (0% H_2), which is obvious to observe that the Pd–N bond remained at 0.14 nm. The characteristic spectra of monomeric Pd appear after bubbling with 6–20% H_2. On increasing the partial pressure of H_2 to 50–100%, an intense Pd–Pd peak appears at 0.25 nm (phase shift uncorrected), which is characteristic of agglomerated Pd^0. Therefore, it could be noted that the formation of monomeric Pd is realized under the H_2 pressure of 6–20%.

8.3 Formation of the Atomically Dispersed Pd⁰ Through H₂ Bubbling in o-Xylene 161

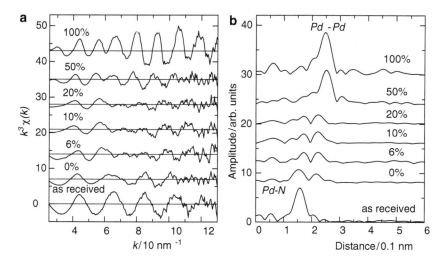

Fig. 8.14 Pd K-edge EXAFS $k^3\chi(k)$ (**a**) and their Fourier transforms (**b**) of Pd(NH$_3$)$_4$Cl$_2$/USY measured after bubbling with 0–100% H$_2$ at 383 K

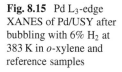

Fig. 8.15 Pd L$_3$-edge XANES of Pd/USY after bubbling with 6% H$_2$ at 383 K in o-xylene and reference samples

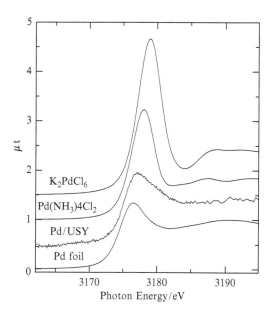

In order to obtain an insight into the valence state of Pd, data of Pd L$_3$-edge XANES are collected at SR center in Ritsumeikan University at Shiga prefecture, Japan. The measurements are carried out at room temperature while the sample is immersed in o-xylene. Figure 8.15 shows the Pd L$_3$-edge XANES of reference samples together with atomically dispersed Pd. The white line represents the electron transition of 2p$_{3/2}$ to 4d orbital of Pd. It can be seen that the larger the valence of the

Pd compounds, the higher the intensity of the peak. It can be seen that the spectrum of atomic-Pd is close to that of Pd foil (metal Pd). Judging from the peak height of the spectrum, the valence of the atomic-Pd is estimated to be $+0.26$, meaning the electronic state of Pd is electron deficient. Thus obtained atomic-Pd will be applied to the Suzuki–Miyaura coupling reactions in Chap. 9 and the catalytic performances are compared with the XAFS data.

References

1. W.A. Weber, B.C. Gates, J. Catal. **180**, 207 (1998)
2. L. Brabec, J. Novákova, J. Mol. Catal. A **166**, 283 (2001)
3. V.S. Gurin, N.P. Petranovskii, N.E. Bogdanchikova, Mater. Sci. Eng. C **C19**, 327 (2002)
4. G. Jacobs, F. Ghadiali, A. Posanu, A. Borgna, X. Alvarez, D.E. Aresasco, Appl. Catal. A **188**, 79 (1999)
5. B. Wen, Q. Sun, W.M.H. Sachtler, J. Catal.**204**, 314 (2001)
6. J. Wang, C. Liu, J. Mol. Catal. A **247**, 199 (2006)
7. H. Yasuda, T. Sato, Y. Yoshimura, Catal. Today **50**, 63 (1999)
8. K. Okumura, K. Nota, K. Yoshida, M. Niwa, J. Catal. **231**, 245 (2005)
9. K. Okumura, H. Matsui, T. Tomiyama, T. Sanada, T. Honma, S. Hirayama, M. Niwa, Chemphyschem **10**, 3129 (2009)

Chapter 9
Catalytic Reaction on the Palladium-Loaded Zeolites

Abstract Palladiums loaded on zeolites are applied to the various catalytic reactions such as total oxidation of hydrocarbons, selective reduction of NO, and organic reactions including Mizoroki–Heck and Suzuki–Miyaura coupling reactions. The catalytic performance of Pd is correlated with structural characteristics as analyzed by XAFS and acid properties of zeolites.

9.1 Combustion of Hydrocarbons Over Pd-Supported Catalysts

Highly active catalysts for the complete oxidation of hydrocarbons, such as methane and volatile organic compounds (VOCs), have been desired from the viewpoint of environmental protection and the energy generation. Numerous studies on the Pd catalyst have been studied because of the high activity of Pd in these reactions [1]. Although previous attention was primarily directed to the investigation on the Pd itself, support is an important component of catalyst that affects the structure and catalysis of Pd. This is because the size, structure, and oxidation state of Pd change significantly depending on the employed supports as mentioned in Chap. 8. In the present study, in order to elucidate how the acid and base property and the structure of support affect the surface oxidation state and reactivity of Pd, Pd is loaded on various kinds of metal oxide and zeolites as support. The surface oxidation state and structure of Pd is elucidated by X-ray photoelectron spectroscopy (XPS) and EXAFS, respectively, and data are correlated with the catalysis and kinetic data. Such a fundamental study on the interaction between metal and support surface may serve to understand one aspect of the metal–support interaction in the supported catalyst. Another important issue is the improvement in the deactivation caused by water vapor that is evolved as a result of combustion and is included in the air. The problem is solved by two methods: deposition of silica layer on the alumina support and minimizing Al concentration of zeolite supports.

9.1.1 Toluene Combustion [2]

Toluene is widely used as a solvent for paints. Total combustion is an efficient and clean way to eliminate toluene present in the gas phase that is harmful for human life. Toluene combustion reaction is carried out over Pd loaded on various kinds of metal oxides, in order to elucidate how the acid and base property of support affect the surface oxidation state and reactivity of Pd. Figure 9.1 shows the dependence of toluene conversion on the reaction temperature. Pd loaded on MgO and WO_3, which has strong basic and acidic character, respectively, are relatively inactive to the reaction, while metal oxides with weak acid–base property, such as Al_2O_3, SiO_2, and Nb_2O_5, exhibit higher activities. As an exception, Pd loaded on ZrO_2 exhibits the highest activity among all the tested samples.

It is expected that the acid or base property of supports induces a change in surface oxidation state of Pd through an electronic interaction between Pd and support surface, so that the combustion activity changes. Figure 9.2 shows the Pd $3d_{5/2}$ peak position of XPS plotted against the electronegativity of metal cation of support oxide. The samples are thermally treated at 573 K under the oxygen flow prior to the measurement. The electronegativity is regarded as an indication of acid and base property of metal oxide. Dotted lines are peak positions that correspond to the metal Pd or PdO. A linear relationship is obtained between these two parameters on WO_3, Nb_2O_5, SnO_2, Al_2O_3, and MgO, where the peak position shifts toward higher binding energy with increase in the acid property of support, and it becomes close to the position of PdO in Pd/WO_3. The fact means that the surface of Pd is readily oxidized when Pd is supported on the acidic support, whereas Pd supported on the basic support is difficult to be oxidized. The observation can be explained through the electronic character of support, which is related with the acid–base property of metal oxide. That is to say, the acidic support with electrophilic character results in

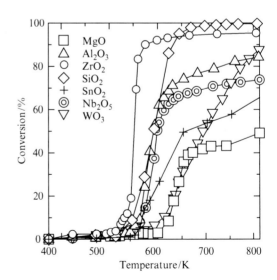

Fig. 9.1 Catalytic combustion of toluene over Pd loaded on various metal oxides

9.1 Combustion of Hydrocarbons Over Pd-Supported Catalysts

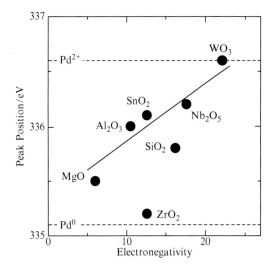

Fig. 9.2 Correlation between maximum Pd $3d_{5/2}$ peak position and electronegativity of metal cation of the support metal oxide

the electron deficient state of Pd, thus the Pd surface is easily oxidized to generate the surface PdO. On the other hand, the basic supports with electrophobic character make Pd particle electron sufficient. Thus the Pd surface becomes difficult to be oxidized. Similar conclusions are obtained in the data of reaction order for oxygen. On the other hand, Pd loaded on SiO_2 and ZrO_2 exhibits exceptional tendency from above relationship in which metallic Pd is relatively stable. In particular, metal Pd is extremely stable on ZrO_2 support. The catalyst exhibits the highest activity in the toluene combustion reaction. Acidity is not detected on these samples according to the temperature-programmed desorption of ammonia. The fact suggests the absence of metal–support interaction between Pd and ZrO_2 surface, since the acidity is responsible for alteration of the oxidation state of Pd as described above. Probably, the high activity in the toluene combustion over Pd/ZrO_2 is due to the generation of metal Pd.

9.1.2 Methane Combustion

Total combustion of methane is important in the field of energy generation because methane is a main component of natural gas. In addition, the evolution of CO_2 is low in comparison with other hydrocarbons, which is beneficial to reduce the emission of CO_2. Although alumina has been frequently employed as the support for Pd, it is expected that the catalytic activity of Pd changes significantly depending on the kinds of supports. Figure 9.3 shows the time course change in the catalytic activity of Pd loaded on three kinds of supports. It can be seen that the behavior depends on the kinds of supports; the catalytic activity of the Pd/Al_2O_3 increases at the initial stage, attains the maximum at ca. 3 h of the elapsed time, and then decreases with

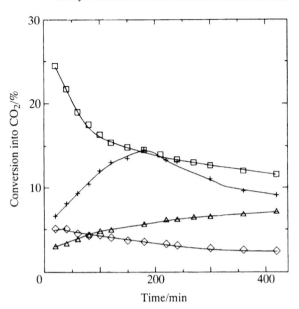

Fig. 9.3 Change of the catalytic activity with time-on-stream on 0.5 wt% – Pd supported on SiO$_2$ (*open square*), Al$_2$O$_3$ (*plus sign*), SiO$_2$–Al$_2$O$_3$ (*open diamond*), and H-mordenite (*open triangle*). Reproduced with permission from [3]. Copyright 1996 Elsevier

duration of time. The catalytic activity of Pd/SiO$_2$ and Pd/SiO$_2$–Al$_2$O$_3$ decreases gradually with time-on-stream at 723 K; the activity is stabilized at ca. 7 h after the initiation of reaction. The stabilized activity of the Pd is ordered in the sequence; on SiO$_2$ > on Al$_2$O$_3$ > on SiO$_2$–Al$_2$O$_3$. The turnover frequency (TOF) of methane oxidation is about 50 times greater than that of Pt supported catalysts, and the activity sequence of support is opposite to that of Pt [4].

Despite the preceding numerous studies on the Pd catalyst, the assignment of active Pd species for the total oxidation of methane is rather controversy; PdO and the mixture of metal Pd and PdO are proposed to be active species. In order to determine the most active Pd phase, methane combustion is carried out over Pd loaded on zeolites of MFI and MOR structure with H-forms, and the influence of Al concentration and structure of zeolite on the oxidation activity of Pd is studied [5]. The catalytic activity of Pd is significantly dependent on the Al concentration and the structure of zeolite supports. A maximum activity is obtained on H-MOR and H-MFI, when the Si/Al$_2$ ratio is 30 and 200, respectively. The initial activity of Pd/H-MOR is higher than that of Pd/H-MFI in the whole range of Al concentration. In order to reveal the origin of the difference between zeolites with MFI and MOR structures, activation energy, reaction order, and EXAFS spectra are measured. The apparent activation energy of Pd loaded on zeolites with H-MOR and H-MFI structure is calculated to be 131–169 kJ mol^{-1} and 59–63 kJ mol^{-1}, respectively. The values observed on Pd/H-MOR and Pd/H-MFI agree with the previously reported activation energies of metal Pd and PdO, respectively [6]. The reaction order with respect to the oxygen pressure is measured by changing the flow rate of oxygen and N$_2$ balance while holding the methane pressure. Figure 9.4 shows the dependence of catalytic activity on the oxygen pressure. From the figure, the orders for oxygen

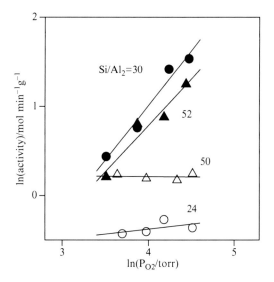

Fig. 9.4 Dependence of the initial methane combustion activity on the partial pressure of oxygen; *filled circle, filled triangle* H-MOR; *open circle, open triangle* H-MFI

on Pd/H-MOR and Pd/H-MFI are calculated to be 1.0–1.2 and 0.1, respectively. The fact indicates that under the practical conditions of catalytic combustion, Pd surface is covered with PdO on MFI structure, since the reaction order for oxygen is zero. In contrast, the first order for oxygen over the Pd/H-MOR means the generation of metal Pd or a mixture of metal Pd and PdO. Therefore, the data of the reaction order for oxygen closely agree with those in the apparent activation energy on both Pd loaded on H-MOR and H-MFI catalysts. The oxidation states of Pd are significantly different on MFI and MOR supports. This fact is directly proven by Pd K-edge EXAFS measured after the reaction shown in Fig. 9.5. The EXAFS spectra were measured at BL01B1 station of Japan Synchrotron Radiation Research Institute (SPring-8) in a transmission mode. In the case of Pd/MFI, Pd keeps the oxidized form even after the reaction. In contrast, a mixture of the metal Pd and PdO phase is found on Pd/MOR. The fact suggests that the partial reduction of Pd occurs during the methane combustion over the Pd/MOR. Taking into account the EXAFS data, a mixture of metal Pd and PdO generated on MOR is attributed to the active species for methane combustion reaction.

Stability of the active Pd species in the hydrothermal conditions is important to keep a high activity in the presence of moisture. This is because under moisture conditions, severe deactivation occurs due to the progressive sintering of PdO via the hydration of PdO to give Pd(OH)$_2$ [7]. The enhancement of thermal stability of the Pd loaded on alumina could be achieved by partial coverage of silica monolayer which is achieved by chemical vapor deposition of silicon alkoxide. The silica monolayer covered on alumina results in not only the enhancement of thermal stability but also the retardation of the sintering of PdO, which is brought by the steric hindrance as illustrated in Fig. 9.6. The optimum loading of deposited silica is determined to be 10 wt%. At this amount, the activity is improved by a factor of two compared with Pd loaded on neat alumina [8].

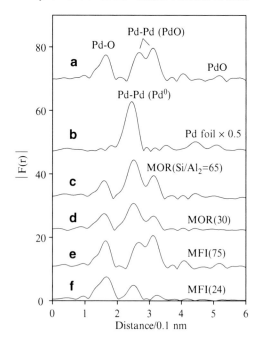

Fig. 9.5 k^3-weighted Pd K-edge EXAFS Fourier transforms for (**a**) bulk PdO, (**b**) Pd foil, (**c, d**) Pd/H-MOR (Si/Al$_2$ = 65, 30), and (**e, f**) Pd/H-MFI (Si/Al$_2$ = 75, 24) measured after methane combustion at 663 K

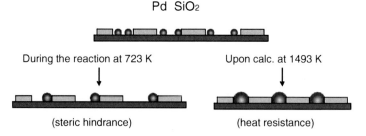

Fig. 9.6 Model of the Pd on 10 wt% SiO$_2$ deposited on Al$_2$O$_3$: change of morphology and sintering of Pd during the reaction and upon calcination at 1,493 K

Another effective way for the enhancement of the stability in hydrothermal condition is achieved by choosing zeolites with a high SiO$_2$/Al$_2$O$_3$ ratio as the support of Pd. Hydrophobicity of zeolite depends on the concentration of Al; zeolites with high SiO$_2$/Al$_2$O$_3$ ratio have the hydrophobic character [9]. As displayed in Fig. 9.7, Pd loaded on the H-β zeolite having the lowest concentration of Al exhibits the most striking performance in that it shows the superior durability and the highest activity in the presence of 10% water vapor. In marked contrast to the zeolite-supported Pd catalysts, Pd loaded on SiO$_2$ is substantially inactive under the same reaction conditions. Based on the Pd K-edge EXAFS coupled with TG analysis, the reason for the superior nature of the high silica H-β is attributed to the hydrophobic character of supports and the formation of slightly agglomerated PdO [10].

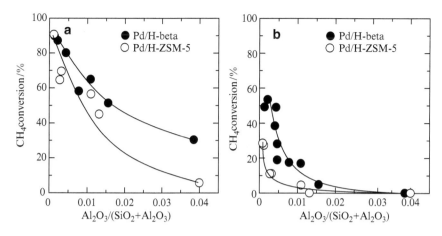

Fig. 9.7 Conversion of methane plotted as a function of $Al_2O_3/(SiO_2 + Al_2O_3)$ of H-beta and H-ZSM-5 measured under the dry conditions (**a**) or in the presence of 10% water vapor (**b**)

9.2 Selective Reduction of NO with Methane in the Presence of Oxygen [11]

For the period of several decades, much attention has been directed to the metal-loaded zeolite catalyst for catalytic reduction of NO_x using hydrocarbons as a reductant. Probably, methane is most preferable because it is the main component of natural gas among hydrocarbons. As for catalyst component, metal elements such as Ga, In, Co, and Pd loaded on zeolites are found to show the high activity and selectivity for the catalytic reduction of NO [12–15]. However, the catalytic activity is sometimes suppressed by water vapor. Therefore, the enhancement of tolerance to water vapor is an important subject. Pd is promising at this point because it is relatively tolerant to water vapor, but the catalytic behavior of Pd is highly dependent on the kinds of the zeolites used as support for Pd. In particular, the existence of Brønsted acid sites is essential to obtain the selective catalyst for reduction of NO. It is widely believed that the acid sites of zeolite keep the Pd^{2+} cation, and this is assigned to the active center for the $NO-CH_4-O_2$ reaction [16–18]. Although the kind of active species of Pd as well as the reaction mechanism is investigated mainly through the observation of NO as a probe molecule [19], the genesis and structure of the active Pd species and the role of the acid sites associated with Pd are still matters of discussion. Sachtler et al. report that the proton in ZSM-5 reacts with PdO to form Pd^{2+} cation [20]. On the other hand, Bell et al. propose a model structure of $Z^-H^+(PdO)H^+Z^-$, where PdO is stabilized on the two adjacent acid sites [21]. To elucidate the structure and nature of active center for selective reduction of NO or the role of acid sites of support, it is necessary to directly observe Pd itself. XAFS is a promising method to achieve a precise determination of Pd structure for metal element dispersed inside zeolite pore systems as described in Chap. 8. As a result, it is inferred that the role of acid sites of zeolite support is to keep the highly dispersed form of PdO, which is active in selective catalytic reduction (SCR) of NO (9.1).

$$CH_4 + 2NO + O_2 \rightarrow N_2 + CO_2 + 2H_2O \qquad (9.1)$$

Unlike the highly dispersed PdO, total oxidation of methane preferentially takes place over the agglomerated PdO according to Scheme 9.2.

$$CH_4 + 2O_2 \rightarrow CO_2 + 2H_2O \qquad (9.2)$$

Taking into account the knowledge obtained by XAFS analysis of zeolite-supported catalysts, we have employed heteropoly acids (HPAs) as an alternative supports for Pd, since HPAs possess strong Brønsted acid character.

9.2.1 Improvement in the Activity Derived by the Combined Effect of Adsorbent of Aromatic Acids [22]

Catalytic reduction of NO using hydrocarbons has been studied as an effective way to convert NO to N_2. In previous studies, numerous kinds of catalysts such as Cu-ZSM-5 and Pt/Al_2O_3 have been found to exhibit a high activity in the reaction [23]. In particular, many researchers have used Pd for catalytic reactions with various reductants using hydrocarbons, alcohols, CO, H_2, and so on [24–27]. On the other hand, the adsorption of NO has been studied as an alternative method of eliminating NO in the gas phase [28, 29]. Several research groups employ Keggin-type HPAs as the adsorbent of NO for this purpose [30, 31]. The adsorption of NO occurs via replacement with the structural water present between the Keggin units of HPAs. The adsorbed NO is activated by O_2 gas with H^+ to yield H^+NO_2. Reversible elimination of NO could also be easily performed by heating the sample above 573 K. We have found that a selective reduction of NO is possible over Pd loaded on HPW/SiO_2 using methane as a reductant [32]. The reaction proceeds at 523 K, which is a significantly lower temperature than that required by the conventional catalysts. Even in the presence of 10% water vapor, the catalyst activity remains unchanged. In the case of this catalyst, it is necessary to combine Pd, HPW, and SiO_2 to obtain the NO reduction. The reaction is expected to take place between NO adsorbed in the acid sites of HPW dispersed over the SiO_2 support and methane dispersed over the Pd center. SCR of NO is performed over Pd/HPW/SiO_2 with various hydrocarbons and alcohols as reductants in order to enhance the possibilities of using the catalyst. Pd/HPW/SiO_2 is physically mixed with zeolites, and the mixture is used in the NO reduction with toluene as the reductant. It is expected that mixing the respective adsorbents for hydrocarbons and NO would lead to a cooperative effect; the catalytic reaction would be promoted by a combination of the activation of NO over HPW and toluene entrapped in zeolites as displayed in Fig. 9.8 [33, 34]. To achieve this, toluene is used as the reductant for NO, and it is found to be effective in the SCR of NO at a relatively low temperature. During the initial period when a vehicle is started, toluene becomes exhausted (cold start period). Hence, it is expected that the removal of toluene and NO from the exhaust gas can

9.2 Selective Reduction of NO with Methane in the Presence of Oxygen

Fig. 9.8 Proposed NO reduction mechanism over the mixed catalyst

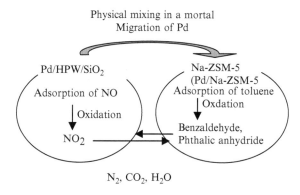

Fig. 9.9 NO, toluene conversion and N_2 yield over Pd/HPW/SiO$_2$ (0.1 g) mixed with various materials (0.1 g) for 30 min. NO, 500 ppm; toluene, 1,000 ppm; O$_2$, 5%; water vapor, 10%; total flow rate, 200 ml min^{-1}; temperature, 523 K

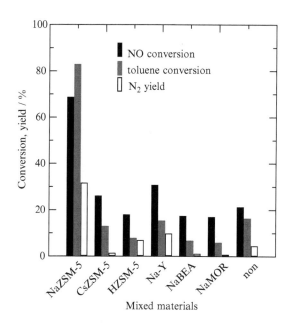

be achieved simultaneously by using a combination of zeolite and HPW; the effect of this combination is demonstrated by our study.

Pd is loaded on the dispersed H$_3$PW$_{12}$O$_{40}$ (HPW) over the SiO$_2$ surface, and the catalyst is applied to the selective reduction of NO with aromatic hydrocarbons. The catalyst exhibits a high activity in the NO reduction when branched aromatic hydrocarbons, such as toluene and xylene, are used as reductants. The catalytic activity of Pd/HPW/SiO$_2$ is improved remarkably by physically mixing it with Na-ZSM-5 as shown in Fig. 9.9. From the TPD of toluene and the analysis of the products, it is inferred that the activity is enhanced when Pd/HPW/SiO$_2$ and Na-ZSM-5 are mixed together. In other words, aromatic hydrocarbons are partially oxidized to yield oxygenated hydrocarbons, e.g., benzaldehyde and phthalic anhydride, over Pd/Na-ZSM-5; in this reaction, a part of Pd migrates from

Pd/HPW/SiO$_2$ to Na-ZSM-5 during the course of the physical mixing procedure. Subsequently, the oxygenated hydrocarbons react with NO entrapped with HPW over Pd to yield N$_2$.

9.3 Cross-Coupling Reactions Over Pd Loaded on FAU-Type Zeolites

Supported metal catalysts have attracted considerable attention due to their characteristics in versatile reactions. This is because well-dispersed metals having surface atoms with a low coordination number are expected to exhibit high activity; this behavior is different from that of a bulk-type catalyst. The formation of a well-dispersed metal has been realized through the reduction of metal precursors, which are protected with stabilizing agents such as dendrimers, micelles, and polymers. However, in general, the formation of metal clusters smaller than subnanometer size is difficult owing to the facile aggregation of metal atoms to give larger clusters or particles. Despite this difficulty, the evolution of high catalytic activity is expected if the highly dispersed metal is generated on a support that is smaller than nanometer-sized clusters. As described in Chap. 8, Pd clusters composed of 13 atoms are readily obtained inside the supercage of USY-zeolite through the exposure of H$_2$ to Pd(NH$_3$)$_4$Cl$_2$/USY at room temperature. The size of the Pd clusters could be regulated by changing the times of O$_2$–H$_2$ exposure. Moreover, formation of the atomically dispersed metal Pd is realized through bubbling with 6%-H$_2$ in o-xylene. The Pd$_{13}$ clusters and atomically dispersed Pd are applied to the Heck and Suzuki–Miyaura cross-coupling reactions, respectively. Thus far, numerous homogeneous Pd catalysts, such as palladacycles and N-heterocyclic carbenes, have been used in these reactions [35–37]. Supported Pd catalysts can also be used in the coupling reactions, since they can be readily prepared and are rather inexpensive in comparison with Pd complexes. Therefore, Pd has been loaded on various types of supports, including active carbon, zeolites, modified silica, hydroxyapatite, and polymers such as dendrimers and polyethylene glycol [38, 39]. Nevertheless, in general, the catalytic activity of these supported catalysts is much lower than that of homogeneous ones. This difficulty may be overcome by fabricating a highly dispersed Pd^0, because monomeric or dimeric Pd^0 leached in a solution has been proposed to be catalytically active in the Heck and Suzuki–Miyaura reactions [40–48].

9.3.1 Heck Coupling Reactions Over Pd Loaded on H-Y Zeolites [49]

Heck reaction (Scheme 9.1) was carried out in DMAc (N,N-dimethylacetamide) solvent using Pd-supported zeolites as catalysts.

In particular, attention is paid to the influence of pretreatment conditions and the kind of zeolites on the catalytic activity and the elution of Pd. Pd^{2+} ion-exchanged

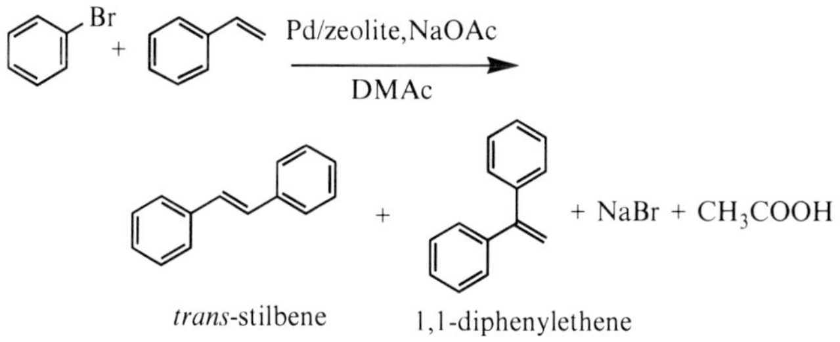

9.3 Cross-Coupling Reactions Over Pd Loaded on FAU-Type Zeolites

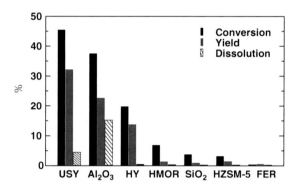

Scheme 9.1 Heck coupling reaction between bromobenzene and styrene

Fig. 9.10 Conversion of bromobenzene, yield of *trans*-stilbene and dissolution of Pd over 0.4 wt% Pd^0 supported zeolite catalysts. Temperature, 393 K; reaction time, 4 h

zeolites exhibit relatively high activity in the reaction. However, considerable amount of Pd is dissolved in the solvent. Deposition of agglomerated Pd^0 is observed after the reaction, suggesting the dissolved Pd^{2+} species is reduced with DMAc during reaction. Pd^0-loaded zeolites are prepared by the reduction of Pd oxide/zeolites with H$_2$. Among Pd^0-loaded catalysts, Pd^0/H-Y is found to exhibit high activity as compared in Fig. 9.10. The dissolution of Pd is significantly suppressed over Pd^0/H-Y as evidenced by ICP analysis of the solution. The recycle use of Pd^0/H-Y is possible through the oxidation and successive reduction with H$_2$ (Fig. 9.11). The growing process of Pd clusters in H-Y is followed by means of energy-dispersive XAFS (partly Quick XAFS) measured during temperature-programmed reduction in diluted H$_2$. It is found that stable Pd$_{13}$ clusters interacted with Brønsted acid sites are generated in the pore of H-Y. The Pd$_{13}$ clusters are ascribed to the active and insoluble species in the Heck reaction.

9.3.2 Remarkable Enhancement of Catalytic Activity Induced by the H$_2$ Bubbling in Suzuki–Miyaura Coupling Reactions [50, 51]

Then the atomically dispersed Pd which is fabricated in the supercage of USY zeolite is applied to the Suzuki–Miyaura coupling reactions (Scheme 9.2).

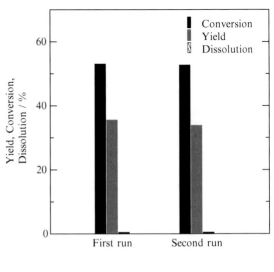

Fig. 9.11 Comparison of the recycle use of the Pd^0/H-Y catalyst in the Heck reaction. Temperature, 393 K; reaction time, 24 h. Pd^0/H-Y used in the second cycle was prepared by the oxidized with O$_2$ at 773 K for 3 h, followed by reduction with 6% H$_2$ at 673 K for 1 h after the first run

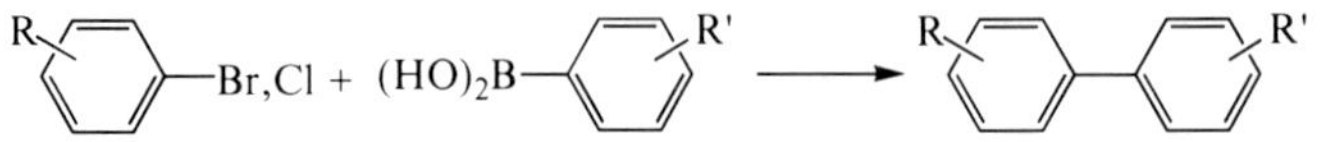

Scheme 9.2 Suzuki–Miyaura coupling reaction between bromo, chloroaryl and phenylboronic acid derivatives

The formation of atomic-Pd is realized through H$_2$-bubbling in o-xylene as described in Chap. 8. It should be emphasized that an extremely small amount of catalyst (0.5–1.0 mg) is employed for the reaction, which corresponded to 1–2 × 10^{-5} mol% with respect to bromobenzene. In order to achieve the in situ formation of the atomic Pd, H$_2$ bubbling is applied during the reaction and the pretreatment period. As a result, the H$_2$ bubbling is effective to enhance the catalytic performance of Pd/USY catalyst to a great extent (Fig. 9.12). Figure 9.13 shows the TOF plotted as a function of the partial pressure of H$_2$. As can be seen in the figure, the addition of only 0.5% H$_2$ to Ar is effective at enhancing the catalytic activity of Pd/USY significantly. Moreover, in good agreement with the in situ EXAFS data given in Fig. 9.15, the TOF largely depends on the H$_2$ partial pressure. The highest catalytic activity is attained at an H$_2$ pressure of 6%, in which the formation of monomeric Pd^0 is observed. The activities of Pd/USY treated with 0% and 50% H$_2$ are much lower, where incompletely reduced Pd species and Pd^0 clusters composed of ca. 13 Pd atoms are observed by in situ XAFS, respectively. Table 9.1 (entries 1, 3–8) lists the results of reactions carried out in the presence of Pd/USY using various derivatives of bromobenzene or chlorobenzene derivatives under 6%-H$_2$ bubbling. Very high turnover numbers (TONs) of up to 13,000,000 are obtained with various substrates in several hours, where the cross-coupling reaction proceeds quantitatively. The catalyst is applicable to the reactions using bulky molecules such as naphthalene derivatives. The reaction between bromobenzene and phenylboronic acid is also performed under an atmosphere of 6% H$_2$. That is to say, a 6% H$_2$ flow is admitted at

9.3 Cross-Coupling Reactions Over Pd Loaded on FAU-Type Zeolites

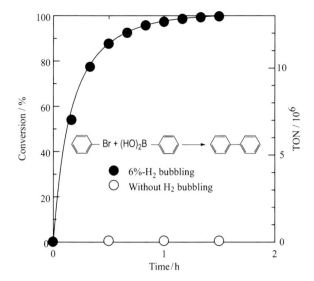

Fig. 9.12 Time course change in the turnover numbers over 0.4 wt%-Pd/USY in the reaction between bromobenzene and phenylboronic acid. *Closed symbols*: with 6%-H$_2$ bubbling; *Open symbols*: without 6% H$_2$ bubbling

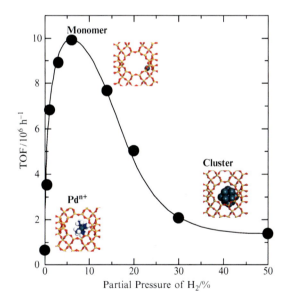

Fig. 9.13 Turnover numbers plotted as a function of the partial pressure of H$_2$ in the reaction between bromobenzene and phenylboronic acid

the upper end of the flask to keep the atmosphere in the flask at 6% H$_2$. The reaction over the 6% H$_2$ atmosphere is also effective at enhancing the activity of the Pd/USY, although the TON obtained under a 6% H$_2$ atmosphere (entry 2, TON = 1,000,000) is lower than that obtained under 6% H$_2$ bubbling (entry 1, TON = 13,000,000).

Table 9.1 Catalytic activity of the 0.4 wt%-Pd/USY in the Suzuki–Miyaura reactions, Ar–Br or Ar–Cl + Ph–B(OH)$_2$ → Ar–Ph

Entry	Ar–Br, Cl	Pd conc. (mol%)[a]	Yield (%)	Time (h)	Turnover number
1[b]	C$_6$H$_5$Br	7.6 × 10^{-6}	99	1.5	13,000,000
2[c]	C$_6$H$_5$Br	6.7 × 10^{-5}	67	3	1,000,000
3[b]	4-CH$_3$C$_6$H$_4$Br	8.7 × 10^{-6}	96	3	11,000,000
4[b]	4-CH$_3$COC$_6$H$_4$Br	9.0 × 10^{-6}	99	1.5	11,000,000
5[b]	4-CH$_3$OC$_6$H$_4$Br	1.0 × 10^{-5}	89	6	8,900,000
6[b]	4-CHOC$_6$H$_4$Br	3.3 × 10^{-5}	99	0.5	3,000,000
7[b]	4-NH$_2$C$_6$H$_4$Br	3.5 × 10^{-5}	83	18	2,400,000
8[b]	4-CH$_3$COC$_6$H$_4$Cl	7.5 × 10^{-2}	3	3	40
9[d]	4-CH$_3$COC$_6$H$_4$Cl	4.4 × 10^{-2}	88	1	2,000

[a]mol% with respect to the bromo- or chlorobenzene derivatives
[b]In o-xylene under 6%-H$_2$ bubbling
[c]In o-xylene in an atmosphere of 6% H$_2$
[d]Under the reaction conditions: ArCl (2.5 mmol), Ph-B(OH)$_2$ (4 mmol), catalyst (50 mg), Cs$_2$CO$_3$ (5 mmol), DMF (6 mL), H$_2$O (0.1 ml), 383 K, Ar atmosphere

Unlike the case of bromobenzene derivatives, the catalytic activity is much lower when 4-chloroacetophenone is used as the reactant (entry 8, TON = 40). However, the higher activity is obtained by choosing appropriate conditions for the reaction (entry 9, TON = 2,000).

9.3.3 A Possible Mechanism for the Formation of Active Pd Species in o-Xylene [52]

Two possibilities are considered for the evolution of outstanding catalytic activity of Pd loaded on the USY for the Suzuki–Miyaura reactions. The first one is the formation of mesopore in the USY support that renders the transportation of reactants and products. This hypothesis may be ruled out by taking into account that the generation of mesopore is not found as confirmed by the nitrogen adsorption isotherms. Another possible mechanism is the stabilization of the atomic Pd interacted with strong acid sites in the USY support. It is well known that the USY zeolite has a strong acid site originated from the extra-framework Al accompanied by the steaming of NH$_4$-Y or Na-Y zeolites, as mentioned in Chaps. 3 and 4. As a result of the dealumination, the characteristic stretching band of the O–H group appeared at 3,595 cm^{-1} in IR spectra, which are ascribed to the strong acid sites of USY generated as a result of the dealumination. As described in Chap. 4, the electron withdrawing effect of AlOH^{2+} unit gives rise to evolution of the strong acid sites. In agreement with the assumption, the catalysis of Pd is sensitive to the steaming temperature for the preparation of USY from NH$_4$-Y; the highest activity is attained at 823 K (Fig. 9.14). The amount of the strong acid sites is measured by means of the IRMS-TPD method to correlate with the catalytic activity. Figure 9.15

9.3 Cross-Coupling Reactions Over Pd Loaded on FAU-Type Zeolites

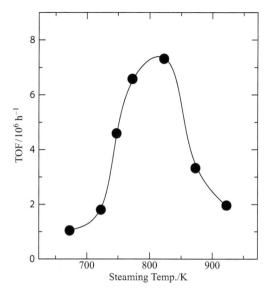

Fig. 9.14 Turnover frequencies plotted as a function of the temperature for preparation of USY zeolite with steaming

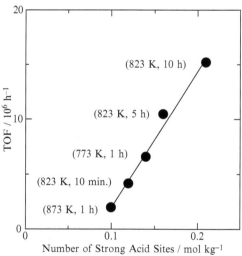

Fig. 9.15 TOF plotted as a function of the number of strong acid sites of USY prepared under different steaming conditions in the Suzuki–Miyaura reaction between bromobenzene and phenylboronic acid. *Numbers in parentheses* indicate the temperature and time for the preparation of USY through the steaming of NH_4-Y zeolite with 18%-water vapor

shows TOF plotted as a function of the number of strong acid sites of USY prepared under different steaming conditions. As can be seen from the figure, a linear relationship is obtained between the amount of the strong acid sites and TOF. The fact means that the strong acid site plays an important role in the evolution of extremely high catalytic activity of Pd. Probably, the atomic Pd is preserved through the interaction between strong acid site and Pd during the reaction, thus sintering is suppressed. Another possible role of the strong acid sites is to tune the electronic properties of Pd. That is to say, it has been reported that the reductive elimination of product (biphenyl) is promoted by the electron withdrawing effect of Pd. The electronic effect of strong acid sites promotes the reductive elimination of products from the active center; thus the reaction is accelerated extensively. Thus, we notice an interesting role of the zeolite acidity; i.e., the interaction of Pd with the strong acid site of zeolite induces the extremely high catalytic activity in the organic reaction.

References

1. Y.H. Chin D.E. Resasco, Catalysis, vol. 14 (Royal Society of Chemistry, London, 1998)
2. K. Okumura, H Tanaka, T. Kobayashi, M. Niwa, Appl. Catal. B 44, 325 (2003)
3. K. Muto, N. Katada, M. Niwa, Appl. Catal. A **134**, 203 (1996)
4. M. Niwa, K. Awano, Y. Murakami, Appl. Catal. 7, 317 (1983)
5. K. Okumura, S. Matsumoto, N. Nishiaki, M. Niwa, Appl. Catal. B **40**, 151 (2003)
6. M. Lyubovsky, L. Pfefferle, Appl. Catal. A **173**, 107 (1998)
7. B. Pommier, P. Gelin, Phys. Chem. Chem. Phys. **1**, 1665 (1999)
8. K. Muto, N. Katada, M. Niwa, Catal. Today **35**, 145 (1997)
9. H. Maeda, Y. Kinoshita, K.R. Reddy, K. Muto, S. Komai, N. Katada, M. Niwa, Appl. Catal. A **163**, 59 (1997)
10. K. Okumura, E. Shinohara, M. Niwa Catal. Today **117**, 577 (2006)
11. K. Okumura, J. Amano, N. Yasunobu, M. Niwa, J. Phys. Chem. B **104**, 1050 (2000)
12. K. Yogo, M. Ishihra, I. Terasaki, E. Kikuchi, Chem. Lett. 229 (1993)
13. T. Tabata, M. Kokitsu, O. Okada, Catal. Lett. **25**, 393 (1994)
14. Y. Li, J.N. Armor, Appl. Catal. B **1**, L31 (1992)
15. M. Suzuki, J. Amano, M. Niwa, Microporous Mesoporous Mater. **21**, 541 (1998)
16. B.J. Adelman, W.M.H. Sachtler, Appl. Catal. B **14**, 1 (1997)
17. C. Descorme, P. Gélin, C. Lécuyer, M. Primet, J. Catal. **177**, 352 (1998)
18. M. Ogura, M. Hayashi, E. Kikuchi, Catal. Today **45**, 139 (1998)
19. A.W. Aylor, L.J. Lobree, J.A. Reimer, A.T. Bell, J. Catal. **172**, 453 (1997)
20. B.J. Adelman, W.M.H. Sachtler, Appl. Catal. B **14**, 1 (1997)
21. L.J. Lobree, A.W. Aylor, J.A. Reimer, A.T. Bell, J. Catal. **181**, 189 (1999)
22. R. Yoshimoto, T. Ninomiya, K. Okumura, M. Niwa, Appl. Catal. B **75**, 175 (2007)
23. M. Iwamoto, H. Hamada, Catal. Today **10**, 57 (1991)
24. R. Burch, D. Ottery, Appl. Catal. B **13**, 105 (1997)
25. R. Burch, P.J. Millington, Catal. Today **26**, 185 (1995)
26. L.F. de Mello, M.A.S. Baldanza, F.B. Noronha, M. Schmal, Catal. Today **85**, 3 (2003)
27. A. Barrera, M. Viniegra, S. Fuentes, G. Diaz, Appl. Catal. B **56**, 279 (2005)
28. K. Eguchi, M. Watabe, S. Ogata, H. Arai, J. Catal. **158**, 420 (1996)
29. M. Machida, S. Ikeda, D. Kurogi, T. Kijima, Appl. Catal. B **35**, 107 (2001)
30. R.T. Yang, N. Chen, Ind. Eng. Chem. Res. **33**, 825 (1994)
31. S. Hodjati, C. Petit, V. Pitchon, A. Kiennemann, J. Catal. **197**, 324 (2001)
32. K. Okumura, R. Yoshimoto, K. Suzuki, M. Niwa, Bull. Chem. Soc. Jpn. **78**, 361 (2005)
33. M. Otremba, W. Zajdel, React. Kinet. Catal. Lett. **51**, 481 (1993)
34. V.R. Choudhary, K.R. Srinivasan, A.P. Singh, Zeolites **10**, 316 (1990)
35. N. Miyaura, A. Suzuki, Chem. Rev. **95**, 2475 (1995)
36. A. Suzuki, J. Organomet. Chem. **57**, 147 (1999)
37. F. Bellina, A. Carpita, R. Rossi, Synthesis-Stuttgart 2419 (2004)
38. F. Alonso, I.P. Beletskaya, M. Yus, Tetrahedron **64**, 3047 (2008)
39. E.A.B. Kantchev, C.J. O'Brien, M.G. Organ, Angew. Chem. Int. Ed. Engl. **46**, 2768 (2007)
40. K. Mori, K. Yamaguchi, T. Hara, T. Mizugaki, K. Ebitani, K. Kaneda, J. Am. Chem. Soc. **124**, 11572 (2002)
41. S. Paul, J.H. Clark, J. Mol. Catal. A **215**, 107 (2004)

References

42. G. Budroni, A. Corma, H. García, A. Primo, J. Catal. **251**, 345 (2007)
43. J.P. Simeone, J.R. Sowa, Tetrahedron **63**, 12646 (2007)
44. N. Jiang, A.J. Ragauskas, Tetrahedron Lett. **47**, 197 (2006)
45. W. Han, C. Liu, Z.-L. Jin, Org. Lett. **9**, 4005 (2007)
46. D.E. Bergbreiter, P.L. Osburn, A. Wilson, E.M. Sink, J. Am. Chem. Soc. **122**, 9058 (2000)
47. G. Durgun, Ö. Aksin, L. Artok, J. Mol. Catal. A **278**, 189 (2007)
48. L. Djakovitch, K. Koehler, J. Am. Chem. Soc. **123**, 5990 (2001)
49. K. Okumura, K. Nota, K. Yoshida, M. Niwa, J. Catal. **231**, 245 (2005)
50. K. Okumura, H. Matsui, T. Tomiyama, T. Sanada, T. Honma, S. Hirayama, M. Niwa, ChemPhysChem **10**, 3129 (2009)
51. K. Okumura, H. Matsui, T. Sanada, M. Arao, T. Honma, S. Hirayama, M. Niwa, J. Catal. **265**, 89 (2009)
52. K. Okumura, T. Tomiyama, S. Okuda, H. Yoshida, M. Niwa, J. Catal. **273**, 156 (2010)

Index

ΔH on various zeolites, 21
ΔS constancy, 31
$\Delta S(desorption)$, 20
$\Delta S(mixing)$, 20

A
Al–O distance, 73
A zeolite, 114
acid-base reactions, 86
acidity, 5
activation energy, 82, 88
additional OH band, 42
adsorption isotherm of nitrogen, 10
adsorption properties, 89
agglomeration of Ga, 97
Al concentration, 166
Al in framework, 24
alkane physical adsorption enthalpy, 84
alkylation of toluene with methanol, 129
Al-MCM-41, 51
aluminum enrichment, 126
amination of phenol, 97
ammonia desorption heat, 45, 83
ammonia IRMS-TPD, 30
ammonia re-adsorption, 16
aromatic compound, 89
aromatic hydrocarbon, 93
atomically dispersed Pd^0, 157

B
Ba, 39, 71
bending vibration of NH_4^+, 32
benzene-filled pore, 108
beta (β) zeolite, 168
bi-molecular mechanism, 80
BLYP, 62
bond lengths and angles, 73

broadly distributed acid strength, 54
Brønsted acid site, 80, 86
Brønsted acid strength, 77

C
Ca, 39, 71
Ca, Ba and La ion exchanged Y, 70
CaNaA, 114
carbenium cation, 80
carbonium cation, 79
carboxylate anion, 110
catalyst, 1
cation exchanged site, 71
cation exchanged Y zeolite, 69
chabazite, 37, 63, 141
chemical liquid deposition (CLD), 112, 136, 138
chemical vapor deposition (CVD), 103, 129
clusters, 150
constancy of ΔS (desorption), 19
constraint cage structure, 37
coordination environment, 74
corrections, IR band position and ΔS, 31
cracking of 1,3,5-triisopropylbenzene, 124, 133
cracking of alkane (paraffin), 79
cross-coupling reactions, 172
curvature of micropore wall, 78, 85
curve fitting of IR- and MS-TPD, 54
curve fitting method for acid strength calculation, 20
curve-fitting analysis of EXAFS, 158

D
$DMol^3$, 62
deactivation, 89
deformation vibration, 58

D (cont.)

desorption enthalpies, 87
detector of ammonia, 12
determination of ΔH, 19
dewaxing, 139
DFT (density functional theory), 61
different four kinds of Brønsted OH, 35
differential changes of the difference spectra, 31
discrimination between Brønsted and Lewis acidities, 34
disproportionation, 132
distorted structure, 47
distribution of ΔH, 55
distribution of Brønsted acid sites, 43
distribution of xylene isomers, 131
distributions of acid strengths, 51
double 6-rings, 35
double numerical plus polarization (DNP) basis set, 62
DXAFS, 149

E

EDTA (ethylenediaminetetraacetic acid)-treated USY, 41, 85
electron acceptor, 40
electron withdrawing nature, 93
electronegativity, 164
electrostatic potential, 71
embedded cluster, 62
enhancement of the Brønsted acidity, 40
enthalpy and entropy changes upon desorption of ammonia, 18
entropy, 91
epitaxial formation, 124
epitaxial growth, 111
equilibrium conditions, 15
equilibrium confirmation in TPD experiment, 31
escape depth, 121
EXAFS, 121
exchange and correlation functional, 62
experimental apparatus of TPD, 11, 30
experimental conditions of ammonia TPD, 13
experimental methods of ammonia TPD, 29
external surface acidity, 134
external surface acidity: measurements, 124
external surface area, 108
extinction coefficients, 34, 55
extra-framework Al, 41, 72
extra-framework Ga species, 97

F

FAU-type zeolite, 154
first-order dependency, 81
formation of a carbonium cation, 88
four kinds of the Brønsted OH bands, 37
framework density, 84
framework topology, 85
frequency of OH stretching vibration, 45
Friedel-Crafts alkylation, 95
fundamental equation of ammonia TPD, 17

G

$Ge(OCH_3)_4$, 106, 121
GeO_2, 121
Ga-containing silicate, 94, 97
Gaussian distribution, 21
generation mechanism of acid site, 24
GGA, 62
GGA-BLYP functional, 65, 68
GGA-HCTH functional, 65, 68
green sustainable chemical processes, 1

H

H_2 bubbling, 157, 174
H_2S, 143
$H_3PW_{12}O_{40}$, 171
h-peak, 13
Haag-Dessau model, 88
HCTH, 62
heat-resisting metal oxide, 143
Heck coupling reactions, 172
heteropoly acids, 170
highly dispersed PdO, 152
hydration of ethene, 55
hydrogenation of acetylene, 144
hydrophobicity, 168
HZSM-5 *in situ* and *ex situ* prepared, 136

I

ICP-ES, 10
Identification, l- and h-peaks, 13
identification of desorbed ammonia, 13, 14
impregnation of Ga on ZSM-5, 98
in situ and *ex situ* prepared H-type zeolites, 26
in situ cell, 154
in situ preparation of H-type zeolite, 26
in situ prepared sample, 26
in situ production, CVD zeolite, 135
inactivation, external surface acidity, 124
IR-TPD, 31, 47
IRMS-TPD experiment, 29

Index

IRMS-TPD measurement, 29
isobutene, 144

K
$k^3\chi(k)$ EXAFS, 156
Keggin-type heteropoly acids, 170

L
l- and h-peaks, 13
l-peak, 13
Lambert–Beer law, 56
LDA-VWN functional, 65
Lewis acid site, 47, 48, 94, 97
linear relationship between DE (alkane cracking) and DH (ammonia desorption), D, delta, 88
loading property, 5
local geometry, 72
long life in phenol amination, 98

M
MAS NMR, 10
material balance of ammonia, 17
Material Studio, 62
maximum number of acid site against the concentration of framework Al, 24
maximum surface concentration, 27
mechanism of CVD, 109
mesoporosity, 83
mesoporous silicas, 93
mesoporous ZSM-5, 47
methane, 169
methane combustion, 165
methanol conversion into gasoline, 131
MFI, 166
MFI, FER, and MWW calculated, 67
microcalorimetry, 10
molar extinction coefficients, 57
molecular imprinting, 111
mono-molecular mechanism, 81
monolayer, 52
monomeric Pd, 160
MOR, 166
mordenite, 44, 153
MS-TPD, 31
Mulliken charge, 74
multivalent metal cation-exchanged Y zeolites, 39

N
Na^+, adsorption of toluene on, 89
Na-ZSM-5, 90, 171

NaA, 114
NH_2 formation, 98
Ni-KA modified, 144
number of acid sites, 24
number of acid sites correlated with Al concentration, 24
number of Na atoms, 92
numerical analysis, 75

O
o-xylene, 113, 157, 174
octane, 139
OH band intensity, 33
origin of acid strength, 77

P
Pd^0/H-Y, 173
Pd_{13} clusters, 173
p-diethylbenzene, 132
p-ethyltoluene, 132
pK_a, 86
Pd, 142, 150
Pd L_3-edge XANES, 161
Pd cluster, 153
PdO, 150
periodic boundary conditions, 62
petroleum refinery, 1
physical adsorption heat, 83
pore-opening size, 134
positions of surrounding atoms, 74
principle of shape selectivity, 129
product selectivity, 103
product shape selectivity, 140
proton donation ability, 46
protonation, 80
PtNaA, 142
pyridine, adsorption of, 10

Q
QXAFS, 149

R
re-adsorption, 91
reactant selectivity, 103
reaction mechanisms, 80
reference catalyst, 126
reference spectra, 30
Rho, 141
Ritsumeikan University, 161
role of cation for enhancement of Bronsted acidity, 40

S

SbCl$_5$, 138
SiCl$_4$, 138
SiO$_2$, 165
Si(OC$_2$H$_5$)$_4$, 135
Si–O$_a$–Al angle, 74
SCR, 170
selective cracking of linear alkane, 139
selective reduction of NO, 151, 169
selective reduction of NO with methane, 142
SEM, 9
shape selectivity, 5, 103, 129
shift of IR band position, 72
Si–O bond distance, 63
Si–O–Si bond angle, 64
SiCuHZSM-5, 144
silane, 105
silica alumina catalyst, 51
silica monolayer, 167
silica overlayer, 111
silicalite, 113, 123
silicon tetra-alkoxide, 104
simulated spectrum, 18, 21
sodalite cage, 35, 70
SPring-8, 149, 167
SR center, 161
SSZ-13, 63
standard entropy of desorption, 87
steaming, 176
steaming of Y zeolite, 71
strain around acid site, 85
strength of acid site, 16
strength of Brønsted acid site, 98
stretching frequency, 45
strong acid sites, 176
supercage, 154
surface density of acid site, 27
Suzuki-Miyaura cross coupling reactions, 172
synthesis method of mesoporous H-ZSM-5, 50

T

TEM observation, 123
temperature of ammonia desorption, 16
temperature-programmed desorption (TPD) of ammonia, 11
tetraethoxysilane, 135
theory for ammonia TPD, 15
thermodynamic description, 86
thermodynamics controlled products distribution, 131
thickness of silica overlayer, 120
TiCl$_4$, 138
TOF, 82, 174
toluene combustion, 164
toluene desorption, heat of, 92
TONs, 174
TPD cell, 11
TPD of ammonia, 15
TPD of toluene, 90
TPO, 156
TPR, 156
triisopropylbenzene, 119
trimethylpentane, 139

U

Ultrastable Y (USY) zeolite, 41, 79, 119

V

V$_2$O$_5$/TiO$_2$, 52
volatile organic compounds, 163

W

WO$_3$/TiO$_2$, 55
WO$_3$/ZrO$_2$, 52
water vapor treatment, 14

X

X-ray absorption fine structure (XAFS), 149
X-ray photoelectron spectroscopy (XPS), 10, 164
XPS measurements, 120
XRD, 9

Y

Y zeolite, 14, 35

Z

ZrO$_2$, 165
zeolite, 1
zeolite framework topology, 77
ZK-5, 141
ZSM-5, 151